# Web Service Mining

George Zheng • Athman Bouguettaya

# Web Service Mining

## Application to Discoveries of Biological Pathways

Foreword by Boualem Benatallah

 Springer

George Zheng
Science Applications International Corporation
Systems & Technology Solutions
7990 Science Applications Ct.
Vienna, VA 22182
USA
GEORGE.ZHENG@saic.com

Athman Bouguettaya
CSIRO ICT Center
Computer Science & Information
Technology
Australian National University
Canberra, ACT 2601
Australia
athman.bouguettaya@csiro.au

ISBN 978-1-4899-8959-8    ISBN 978-1-4419-6539-4 (eBook)
DOI 10.1007/978-1-4419-6539-4
Springer New York Dordrecht Heidelberg London



Printed on acid-free paper

Springer is part of Springer Science+Business Media (www.springer.com)

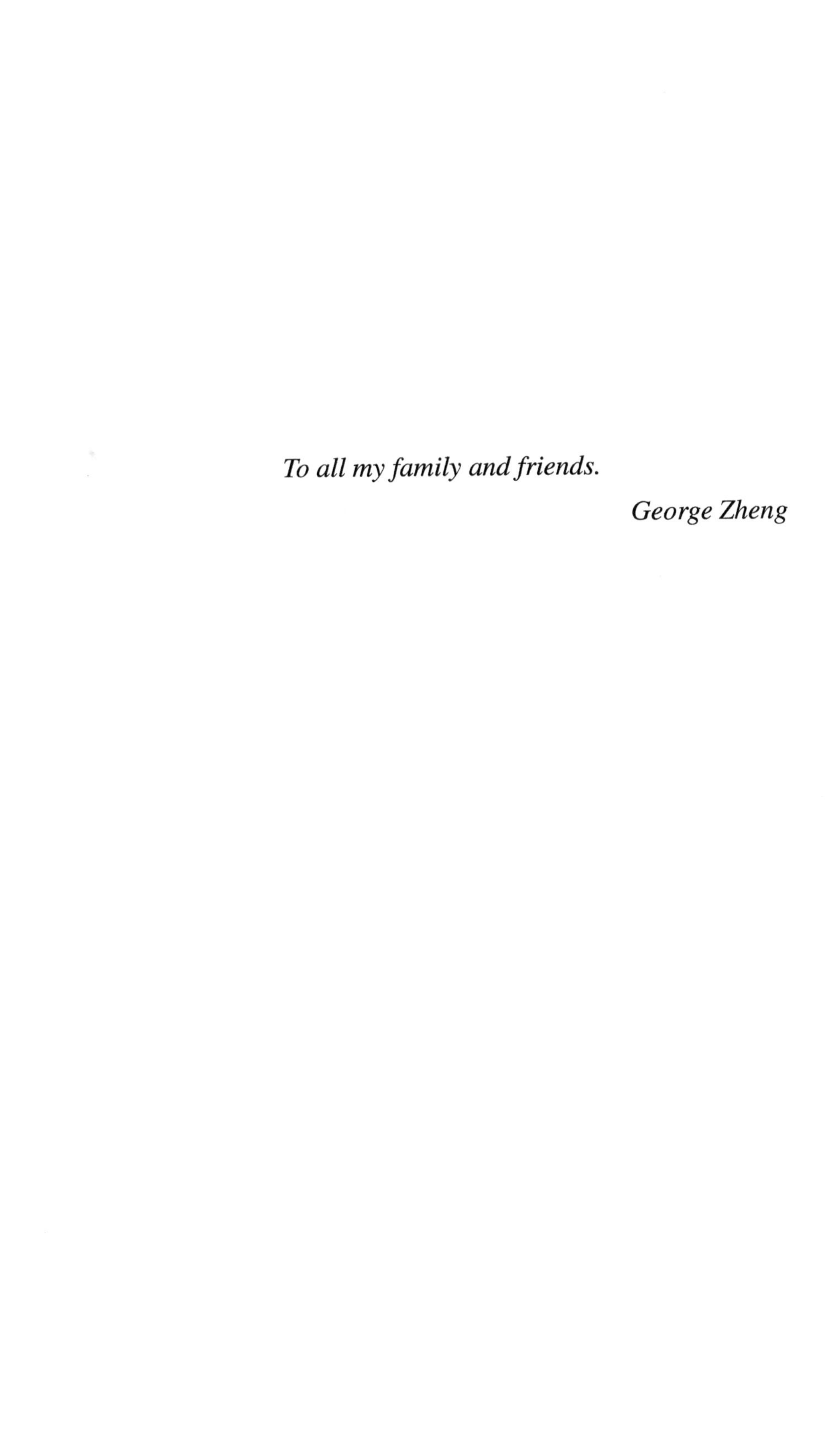

*To all my family and friends.*

*George Zheng*

*To Malika.*

*Athman Bouguettaya*

# Foreword

The new computing environment enabled by advances in service oriented architectures, mashups, and cloud computing will consist of service spaces comprising data, applications, infrastructure resources distributed over the Web. This environment embraces a holistic paradigm in which users, services, and resources establish on-demand interactions, possibly in real-time, to realise useful experiences. Such interactions obtain relevant services that are targeted to the time and place of the user requesting the service and to the device used to access it. The benefit of such environment originates from the added value generated by the possible interactions in a large scale rather than by the capabilities of its individual components separately. This offers tremendous automation opportunities in a variety of application domains including execution of forecasting, office tasks, travel support, intelligent information gathering and analysis, environment monitoring, healthcare, e-business, community based systems, e-science and e-government.

A key feature of this environment is the ability to dynamically compose services to realise user tasks. While recent advances in service discovery, composition and Semantic Web technologies contribute necessary first steps to facilitate this task, the benefits of composition are still limited to take advantages of large-scale ubiquitous environments. The main stream composition techniques and technologies rely on human understanding and manual programming to compose and aggregate services. Recent advances improve composition by leveraging search technologies and flow-based composition languages as in mashups and process-centric service composition.

The authors provide a very informative discussion on the state of the art in service mining. The motivation is simple: existing service composition techniques are only appropriate when a service composer knows how to search effectively for required services and have the programming skills to perform composition using flow based languages or general purpose programming language. The approach proposed by the authors rely on discovering useful and interesting services that can be used in situations where users lack precise knowledge to perform targeted search of services to be composed. The realization of the service mining approach poses a difficult challenge to address the critical issue of effort-less and flexible exploration and manipulation of services in autonomous and highly evolving service spaces. The authors provide the fundamental concepts behind Web service mining and illustrate the proposed concepts and techniques in application related to biological pathway discovery. The book is timely, provides a thorough scientific investigation and also has practical relevance in the general area of composition. The book will be of particular interest to researchers and professionals wishing to learn about relevant concepts and techniques in service composition and mining.

Sydney, Australia                                        *Boualem Benatallah*

December 2009

# Preface

The ubiquity of the Web is enabling innovations that were thought to be beyond reach just two decades ago. The Web has now in effect become the dominant medium for all human and economic activities. The advent of Web services almost a decade ago has accelerated this trend and included such activities as e-government, e-commerce, and e-science applications. The ultimate goal of this enabling technology is the use of Web services as independent components that are automatically (i.e., without human intervention) formed and that could dissolve or persist post demand-completion.

The fast increasing number of Web services is transforming the Web from a data-oriented repository to a service-oriented repository. Web services are anticipated to form the underlying technology that will realize the envisioned "sea of services". Web services have originally been driven by standardization bodies, thus eliciting a wide acceptance in businesses and governments. The Service-Oriented Architecture (SOA) was conceived as the IT industry response to leveraging the Web as the center of all activities. This was achieved by adapting and evolving such technologies as CORBA to be Web congruous. As aforementioned, the development of Web services has so far mostly been the result of standardization bodies usually operating on a consensus basis and driven by market considerations. In this context, innovation and long-term market effects have not been the primary concerns. The standardization process has so far been very fragmented, leading to competing and potentially incompatible Web service infrastructures that lack a sound foundational framework. To maximize the benefits of this new technology, there is a need to provide a rigorous methodology for specifying, selecting, trusting, optimizing, composing, and mining Web services.

In this book, we focus on service mining. We describe a novel foundational framework that lays out a theoretical underpinning for the emerging field of service mining. We describe a disciplined and systematic framework for the efficient mining of Web services functionalities using non-functional properties, called Quality of Web Service (QoWS) parameters. The key components of this approach revolve around a novel service model that provides a formal abstraction of Web service modeling and organization. We draw inspiration from chemical processes for service organization methodologies, and drug discovery for service mining techniques. We use this inspiration as the target application as a proof-of-concept of the proposed framework and algorithms.

We distinguish between Web service composition and Web service mining in terms of the stated *a priori* goals or lack thereof, respectively. For instance, service composition has traditionally taken a top-down approach. The top-down approach requires a user to provide a goal containing specific search criteria defining the exact service functionality the user expects. Often, the more specific the query and search criteria are, the smaller the search space and more relevant the composition results will be. The specificity of the search criteria would reflect the interest and often knowledge of the service composer about the potential composability of existing Web services. Since the composer is typically only aware of and consequently interested in some specific types of compositions, the scope of such a search is usually very narrow. The top-down approach thus works well only if the service composer clearly knows what to look for and the component Web services needed to compose such services are available. Another view, as taken by service mining, approaches service composition from the bottom-up. It aims at exploring the full potential of the service space without any *a priori* knowledge of what exactly is in it. Instead of starting the search with a specific goal, a service engineer may be interested in discovering interesting and useful service compositions as a result of the search process. For performance reasons, a general goal may be provided at the beginning to scope down the initial search space to a reasonable size. The interesting outcome of service mining is the ability to find useful and unexpected compositions. Thus, unlike the search process in the top-down approach that is strictly driven by the search criteria, the search process in the bottom-up approach is *serendipitous* in nature leading potentially to great innovations.


Virginia, USA                                                                      *George Zheng*
Canberra, Australia                                                    *Athman Bouguettaya*

December 2009

# Acknowledgments

I thank my wife Lin for her love and patience, and my daughter Sydney for the joy and hope she constantly brings. I thank my good friend, Traxon Rachell, for his continuing support, advice and inspiration; Folashade (Sharday) Adeyosoye, a good friend and former colleague, for his help in setting up my testbed environment for the application of this work; and Thien Le, a good friend and colleague, for his tips and interesting conversations. I thank SAIC for providing me with the environment and financial support for completing this work. Finally, I thank my mother's love and dedication to me and to my father, whose continuing battle against terrible diseases in the past few years has given me the courage to get involved with bioinformatics – an unfamiliar research field, and to make it a fertile ground for applying this work.

*George Zheng*

I would like to acknowledge the support of my family during the preparation of this book: my wife Malika, my children: Zakaria, Ayoub, and Mohamed-Islam. I would also like to thank my employer CSIRO (Australia) for providing me the environment to successfully finish this work.

*Athman Bouguettaya*

# Contents

# Chapter 1
# Introduction

The Web was invented in the early 1990's by Tim Berners-Lee to facilitate information sharing among scientists [101]. It wasn't until 1993 when the Web suddenly became widely known among the public, thanks to the release of a Web browser called Mosaic. Mosaic was made available to the average user and quickly captivated the imagination of the public about the Web through its intuitive graphical user interface. Governments, businesses and individuals realized the great potential of the Web and took advantage of its growing popularity by making their data and applications Web accessible. Although started as a repository of information containing merely text and images, the Web has since evolved into a host of both multimedia information and service-providing applications. These applications range from map-finding to weather-reporting to e-commerce. Some of the applications involve, in real-time, hardware devices such as temperature sensors and traffic monitoring cameras. Businesses and governments realized that by integrating existing Web applications, they can provide value-added services that were previously unavailable. Unfortunately, the customized interfaces required to access these applications and the lack of semantics in the data they consume and generate have made the seamless integration of these applications and quick deployment of value-added services a daunting challenge. To address these problems, two enabling technologies have arisen to prominence. The first is the initiatives of *Web service* announced around the year of 2000 by IBM, HP, Sun, Oracle and Microsoft. These initiatives include IBM's Web services, HP's e-speak, Sun's Open Network Environment (ONE), Oracle's network services, and Microsoft's .NET. At the core of all these initiatives is the assumption that companies will in the future buy their information technologies as services provided over the Internet rather than owning and maintaining all their own hardware and software [56]. As a result of collaborative effort among these major companies and other institutions, the World Wide Web Consortium (W3C) then published the specification of Web service [98]. According to this specification, a Web service is a Web application whose functionalities can be programmatically accessed via a set of homogeneous Extensible Markup Language (XML) interfaces. The standard Web service architecture from this specification, as shown in Figure 1.1, allows such an application to be first *described* by one organization (i.e., ser-

G. Zheng and A. Bouguettaya, *Web Service Mining: Application to Discoveries of Biological Pathways*, DOI 10.1007/978-1-4419-6539-4_1,
© Springer Science+Business Media, LLC 2010

vice provider) and *published* to a *service registry*, and then *discovered* and *invoked* later by *independently* developed applications (i.e., service consumers) from other organizations, essentially making these applications interoperable on the Web at the application interface level.

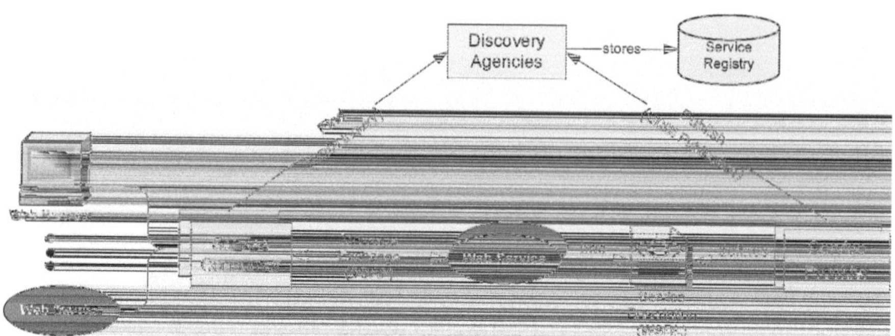


**Fig. 1.1**   Standard Web Service Architecture

The second enabling technology is the use of *ontologies* in describing the content these applications process [50]. Ontologies can be traced back to the Artificial Intelligence (AI) community, where they were proposed to facilitate knowledge sharing and reuse. They have become popular in many other fields as they are increasingly seen as key to enabling semantics-driven data access and processing. An ontology allows the specification of a *shared conceptualization* in an *explicit, machine-readable* format, resulting in a standard way of interpretation. Ontologies make it possible to build a controlled vocabulary of concepts that can be used by the applications to unambiguously describe the content they generate and interpret the content they consume. The reliance on ontologies results in the loosening of the tight semantic coupling that would otherwise be required between a service providing application and service consuming application. This recent trend of empowering Web services with semantics using ontologies has consequently brought about the next generation of Web services called *Semantic Web services*. The Semantic Web service deployment of independently developed applications allows these applications to interoperate on the Web at both the application interface and information semantics levels, making the integration of these applications for value-added services a much easier and more inexpensive task, as compared to earlier approaches that relied on technologies such as Electronic Data Interchange (EDI), Common Object Request Broker Architecture (CORBA) and Enterprise Java Beans (EJB).

The advent of Semantic Web services prompted W3C to revise its Web service architecture [20], as illustrated in Figure 1.2. According to this architecture, the discovery service first obtains (e.g., via search engine or publication from service provider) both the Web Service Description (WSD) in Web Service Description Language (WSDL) [13] and an associated Functional Description (FD) of the provider service (step 1a). The functional description describes the functionality of the provider service. It is machine-processable based on formats such as Resource Description Framework (RDF) [8], DARPA Agent Markup Language for Services (DAML-S), Web Ontology Language for Services (OWL-S) [99] or Web Service Modeling Ontology (WSMO) [11]. The requester then supplies criteria to the discovery service to select a WSD based on its associated FD (steps 1b and 1c). Step 2 ensures that both the requester and the provider *agree* on the semantics of the desired interaction through the use of ontologies. The remaining steps are essentially the same as before.

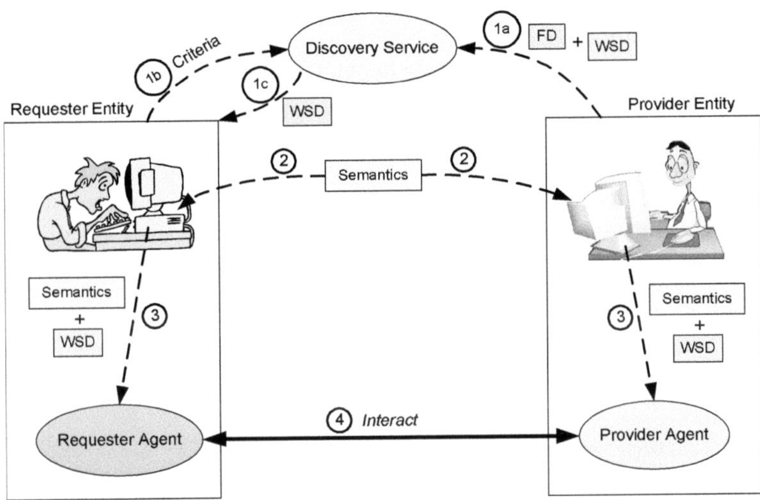


**Fig. 1.2**  Semantic Web Service Architecture

The ease of application integration enabled by the use of Web services and ontologies contributed to the increasing popularity of *Web service composition*, which has attracted governments and businesses as it is identified as a new way of facilitating business-to-business (B2B) and inter-government agency collaborations. Web service composition aims at providing value-added services through composing existing services. Instead of taking high-risk, big-bang approaches by starting from scratch and manually developing a new system every time, businesses and governments can focus initially on opportunities that will yield immediate efficiency gains. As new services become available on the Web, they can then take advantage of these services by assembling them in a way that meets specific service requirements. This new practice would enable both governments and businesses to reduce overhead and deploy their solutions more quickly, mitigating the risk of organizational disruption as they transition into the service architecture. To help standardize service composition, two different types of approaches were independently developed. On the one hand, the business world targeted the earlier Web service paradigm and developed a number of XML-based standards. This include Web Services Flow Language (WSFL) [15, 65], Web Service Choreography Interface (WSCI) [97], XLANG (an XML-based extension of WSDL) [92] and Business Process Execution Language for Web Services (BPEL4WS) [42]. They are largely focused on formalizing Web service specifications, flow composition and execution based on their syntactical characteristics [90]. On the other hand, the Semantic Web research community focused more recently on using the concept of Semantic Web services and developed complementary standards such as DAML-S, OWL-S [99], WSDL with Semantics (WSDL-S) [16] and WSMO [11]. While the two types of standards complement each other, they have both taken a top down approach. The top down approach requires a user to provide a goal containing *specific* search criteria defining the exact service functionality the user expects. An example of such a goal may be to compose a travel service that supports flight booking, car rental and hotel reservation, as illustrated at the top left of Figure 1.3. The corresponding search criteria may state that the travel service shall respond to a runtime search request for a place and a time duration with a list of options containing descriptions of price, luxury, convenience and flexibility for each. Given such a goal, a service composition tool may search for existing component services that individually support some of these four functional areas and try to match them for the same geographic locations.

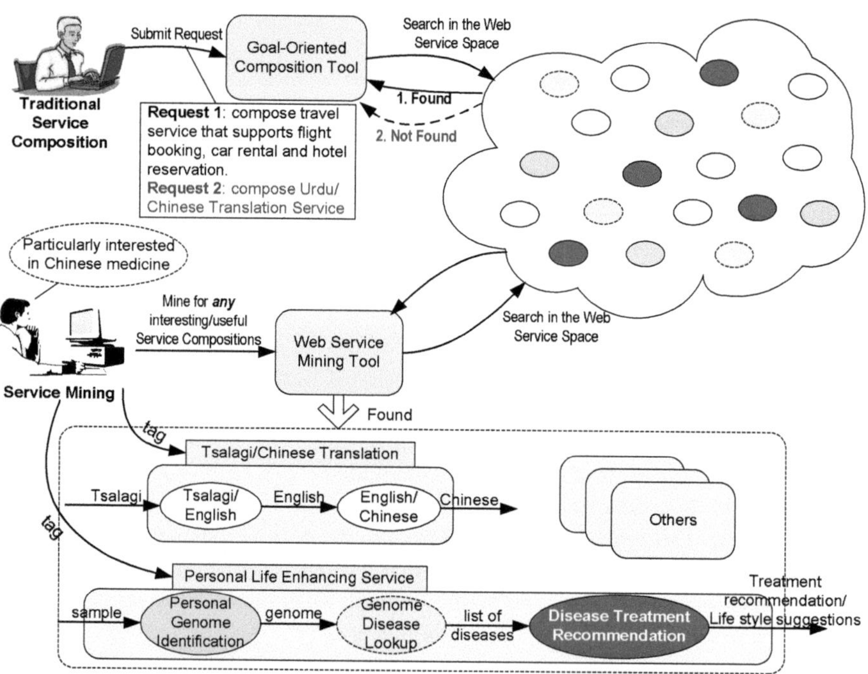


**Fig. 1.3**  Top down Composition vs. Bottom up Mining

Often the more specific the query and search criteria are, the smaller the search space and more relevant the composition results will be. Vague criteria, on the other hand, not only increase the scope of the search space, which can result in longer search time, but also make the relevance of composition results difficult to determine. The specificity of the search criteria would thus reflect the interest and often knowledge of the service composer about the potential composability of existing Web services. Since the composer is typically only aware of and consequently interested in some specific types of compositions, the scope of such a search is usually very narrow. As a result, an attempt to start the search with a set of specific search criteria (e.g., compose Urdu/Chinese translation service in request 2) that are not framed to coincide with the availability of interested Web services will most likely end up empty-handed. Thus, the top down approach can work well only if the composite Web service designer clearly knows what to look for and the component Web services needed to compose such services are available. Aiming at exploring the full potential of the service space without prior knowledge of what exactly are in it, another view that approaches service composition from bottom up is building up recently. Instead of starting the search with a specific goal, a service engineer may be interested in discovering *any* interesting and useful service compositions that may come up in the search process. For performance reasons, a general goal may be provided at the beginning to scope down the initial search space to a reasonable size. For illustration purposes, we show at the bottom of Figure 1.3 that a service engineer sets out to find any interesting and useful services with a general interest in Chinese medicine in mind. What comes out of the search process might be quite surprising. For example, in addition to discovering the possibility of composing a service for translating Tsalagi [1] to Chinese, the engineer also discovers, with the help of a service mining tool, a service composition that takes as input a biological sample from a subject, determines the corresponding genome and the possible diseases the subject is predisposed to, and finally generates a list of treatment recommendations and/or life style suggestions. Thus, unlike the search process in the top down approach that is strictly driven by the search criteria, the search process in the bottom up approach is *serendipitous* in nature, i.e., it has the potential of finding interesting and useful service compositions that are *unexpected*. We propose to use *Web service mining*, a bottom-up Web service composition approach, to address the need for early exposure of opportunities for composing interesting and useful services in the absence of specific user goals.

## 1.1 Defining Web Service Mining

As the Web is poised to transition from a Web of data to a Semantic Web of both data and *services*, where Web services would be the *first-class* objects, there will be increasing opportunities to compose potentially *interesting* and *useful* Web services

---

[1] A language spoken by the Cherokee Indian tribe.

from existing services. Since the collective opportunities of composing services will surpass anyone's imagination, many of these opportunities will be hidden in the Web of available services and unexpected to most people. While we may not sometimes have the specific queries needed to search for them, being able to discover these opportunities early in today's business environment equates to gaining competitive business advantages. For government agencies, doing so also means that citizens receive useful and potentially life-enhancing services in advance. It is thus essential to be able to proactively attempt to discover useful composite services even when the goals are unspecified at the moment, or simply hard to imagine or unknown. Much like the easy access to a glut of data that has provided a fertile ground for data mining research, we expect that the increase in Web services' availability will also spur both the need and opportunities to break new ground on *Web service mining*. We believe that Web service mining would be key to help realize the full potential of the Semantic Web services by leveraging the large investments in applications that have so far operated as non-interoperable silos.

We define Web service mining as a search process aiming at the discovery of *interesting* and *useful* compositions of existing Web services. It complements the top down Web service composition approaches in cases where the goals are unavailable or unknown. Web service mining does not assume *a priori* knowledge about Web services' composability. Instead of using a query to search for a specific type of composition in a top down fashion, Web service mining needs to relies on component Web services themselves to cluster up in a bottom up and exploratory fashion. The end results may be simple service compositions or complex service composition networks that are both interesting and useful. We formally define interesting and useful composition of Web services below:

Given a set of Web services $S$, a set of composability rules $R$, interestingness measures $I$, an interesting composite Web service is a service composed using a subset of rules in $R$ of a subset of Web services in $S$, that exhibits high values for a user selected subset of interestingness measures $I$. A useful composite Web service is established similarly but exhibits high values for a user selected subset of usefulness measures $U$. An interesting and useful service composition network is a subgraph in a service composition network consisting of interesting and useful service compositions, among others, that also asserts a user hypothesis with a certainty exceeding a threshold $t$.


## 1.2 Major Challenges

The growing popularity of the Web and Web services has presented a brand new research topic, namely *Web service mining*, which carries significant importance, as we have identified. An effective mining framework would allow potentially interesting and useful composable Web services to be discovered based on their availability, rather than solely on human intuition or knowledge about their composability. Different from traditional composition approaches, which use user specified goals containing specific criteria to drive the search for matching component Web services in a top down fashion, the lack of specific goals in Web service mining leads to a much broader scope that covers a large space of service composition potentials. The exploration of these potentials would lend Web service mining naturally to being carried out using the *bottom-up* strategy. Consequently we are faced with the following two challenges:

***Combinatorial explosion.*** The first challenge is how to discover composable Web services without specific search criteria in mind. A naive approach would be to conduct an exhaustive search and full-blown composability analysis between any two Web services in the service registry; and once a potentially positive composition from two component services is identified, expand the analysis to allow for more component services in the composition. As the number of registered Web services increases at an accelerating rate, such an approach can quickly become infeasible due to the overwhelming computation resulted from a "combinatorial explosion."

***Evaluation of composition interestingness and usefulness.*** The second challenge is how to determine the *interestingness* and *usefulness* of a composed service identified through mining. In the top down approaches, the determination of *interestingness* and *usefulness* is not a major concern, since the goal provided by the user already implies what type of compositions the user anticipates. In Web service mining, neither interestingness nor usefulness would be so obvious when the composed Web services are discovered without any specific goals. Useful composed services may be contaminated with frivolous or trivial ones that add little or no value. In addition, many of those useful services may have already been known. It will thus be necessary to distinguish interesting and useful composed services from those that are either trivial, already known or useless.

As we look for efficient techniques to address the first challenge of combinatorial explosion, similarities between Web services and *molecules* offer some interesting insights. Web services can be thought of in many ways as similar to molecules in the natural world. Like a molecule, a Web service has both attributes and dynamic behaviors. Like a molecule formed from constituent atoms and/or simpler molecules, a composite service is composed of component services. Under the right conditions, certain molecules can *recognize* each other and form bonds in between. The concept of recognition can be easily extended to the Web service world to help devise mining techniques on Web services. While the outcomes of our research are generic enough to be applicable to a wide range of applications, the similarity between molecules and Web services motivated us to apply our mining framework back to the field of bioinformatics. As a result, the discovery of biological pathways came across naturally as the main application of our mining framework. This essentially makes our research a subfield of *natural computing* [60], which "investigates models and computational techniques inspired by nature and, dually, attempts to understand the world around us in terms of information processing."

In anticipation of the need and opportunities to mine Web services, our research focuses on developing a framework that would facilitate the related mining activities and address the challenges of both combinatorial explosion and the determination of Web service interestingness and usefulness. Our objectives include:

- Identify the type of activities involved in the mining process,
- Develop a mining framework containing effective strategies to streamline and efficient algorithms to automate much of these activities,
- Develop measures that can be used to objectively evaluate the *interestingness* and *usefulness* of the mining results,
- Provide support to user for hypothesis formulation
- Determine strategies for subjectively evaluating the *usefulness* of the mining results, and
- Demonstrate the effectiveness of the framework through a motivating use case of biological pathway discovery.

## 1.3 Our Contributions

We began our research by studying existing Web service composition methodologies. We then identified and addressed the need for a new research area not currently covered by these methodologies, i.e., mining Web services on the Semantic Web. In looking for efficient techniques to carry out the mining activities, we studied how molecular interaction and composition take place in nature and drew analogies between molecules and Web services. These analogies led us to the exploration of drug discovery methodologies, which helped us shape our service mining framework. This framework is essentially based on a *bottom up* approach to identify and verify composable Web services. Unlike traditional service composition approaches, it does not assume any *a priori* knowledge about the composability of these services. The similarity between molecules and Web services also motivated us to apply our mining framework back to the field of bioinformatics. In particular, we examined problems faced by existing biological pathway discovery approaches. We proposed to address these problems by first modeling biological processes as Web services and then applying our service mining framework for the discovery of pathways linking these services. We list below our major contribution from this research:

*A **Web service ontology***: We established a Web service ontology consisting of a set of constructs used to organize Web services. In particular, as an extension to existing service categorizing approaches (i.e., OWL-S and WSMO), we propose to describe shared capabilities of Web services in this ontology through the construct of *operation interface*. This construct allows Web services to declare their role as either a provider or consumer of an operation interface, paving the way for them to plug into one another at the operation level.

*A **molecular analogy***: We explored the similarities between Web services and molecules in the chemical world for the purposes of designing an efficient framework for Web service mining. We derived a set of concepts such as *service and operation recognition*, which led to the establishment of several efficient mining algorithms.

***Web service mining framework***: We proposed a Web service mining framework that is inspired by the drug discovery process. We used an ontology-based publish/subscribe mechanism to convert the traditional top down combinatorial search problem into a process of bottom up service and operation recognition. The framework uses specific semantic constraints to verify the composability of identified composed Web services.

***Interestingness and usefulness evaluation***: We developed a set of measures that can be used to objectively determine the interestingness and usefulness of service compositions discovered via service mining.

***Web service modeling of biological processes***: We proposed to model biological entities and processes as Web services. Two modeling approaches currently dominate the way biological entities and processes are represented. The first approach describes biological entities and processes using free text description and annotations. It inherently lacks the ability to support simulation. The second approach relies on isolated computer models to represent biological processes. As a result, they lack the ability of being discovered and invoked in the global Web environment. Our modeling strategy bridges the gap of these two approaches and made it possible for these processes to be discovered as well as invoked on the Web.

***Discovery and simulation of biological pathways***: We applied our service mining framework to service oriented models of biological processes and demonstrated its effectiveness in the automatic discovery of pathways linking these processes. We developed a set of graph related algorithms identifying interesting pathway segments and helping user with hypothesis formulation that would lead to the identification of useful pathways. We also demonstrated the feasibility of simulating these pathways through invoking service oriented biological models directly on the Web.

## 1.4  Book Organization

The book is organized into seven chapters. The following gives a preview of these chapters.

In Chapter 2, we describe how we categorize Web services using an ontological perspective. We use such categorization as the basis when we later describe in Chapter 3 service and operation recognition, a concept that was derived from a molecular analogy. This analogy led us to look into drug discovery process for efficient strategies when developing our service mining framework, which we present in Chapter 4. In Chapters 4.1 to 4.3, we describe in detail different phases of our service mining framework including algorithms used to automate mining activities during these phases. In Chapter 5, we establish the context of our application and describe the state-of-the-art in computer based pathway discovery. We then introduce our approach as advancement to existing approaches. We describe in detail how biological entities and processes can be modeled using WSDL and semantically exposed via

WSML [10] and corresponding adapters. We show how these services can be deployed into the supporting runtime infrastructure. We then apply our service mining framework to pathway discovery and demonstrate the effectiveness of our framework. In Chapter 6, we describe related research fields including Web service composition, data mining, and log-based interaction and process mining. In addition, we point out main differences between our research topic and these fields. We also describe natural computing, a field of research that our works belong. In Chapter 7, we provide concluding remarks and point out directions for future research.

# Chapter 2
# An Ontological Perspective of Web Services

Web service mining requires Web services, the subject of mining, to be clearly defined so that mining techniques can be used to unambiguously target them. A Web service ontology provides an effective means for describing various Web service related concepts and their relationships. We represent the Web service ontology using a Unified Modeling Language (UML) diagram in Figure 2.1. In the following, we define key concepts depicted in the diagram.

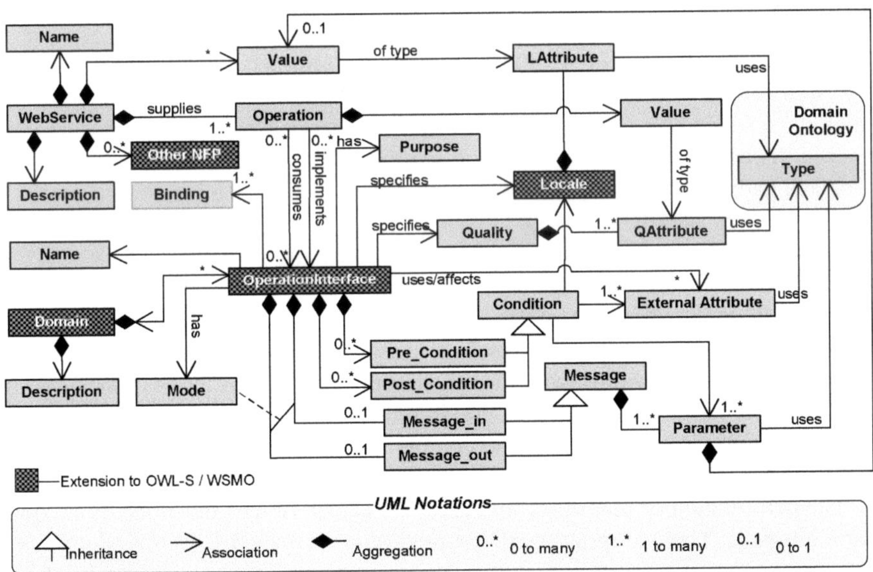


**Fig. 2.1** Web Service Ontology


G. Zheng and A. Bouguettaya, *Web Service Mining: Application to Discoveries*
*of Biological Pathways*, DOI 10.1007/978-1-4419-6539-4_2,

***Message***. Web services communicate with each other through the exchange of XML messages. A message consists of one or more parameters. A parameter has a value of a certain data type. There are two types of messages: $Message_{in}$ and $Message_{out}$. $Message_{in}$ is used to send parameters to a service operation, $Message_{out}$ is used to return results from a service operation. $\Diamond$

***Condition***. A condition can be either a pre-condition or a post-condition. A pre-condition specifies the necessary condition for an operation to be activated. This may be expected values in $Message_{in}$ or relationships between these values. A post-condition describes the state of the information space after the execution of a service operation. This may be related to parameters in $Message_{out}$. $\Diamond$

An example of a pre-condition would be that the quantity or kinetic energy of an input parameter (or substance) to a biological process has to exceed a certain threshold. An example of a post-condition may be the dissipation condition that must be associated with an output parameter (or substance) generated by a biological process.

***Domain***. Web service operations are categorized by domains of interest. A domain contains a description that describes the purpose of the domain and summarizes its functionalities. $\Diamond$

Examples of domains include *travel, home entertainment, attorney services, credit check, healthcare, drug design,* etc.

***Locale of Interest***. Locale of interest $L = \{l\}$, where each $l$ represents a semantic restriction containing attributes that describe the applicability of an operation interface. $L$ can be used to limit the scope of mining. $\Diamond$

An example of semantic restriction would be a regional or geographical boundary. Web service operations are sometimes applicable only to certain regions. For example, a *reserveCar()* operation of a car rental service may require the car to be used in the continental US. $L$ is used to specify such restrictions. In the context of drug and pathway discovery, an $l$ may specify the locality of a certain receptor.

***Quality of Web Service (QoWS)***. QoWS $Q = \{q\}$ is used to evaluate an operation provided by a Web service, where $q$ is a quality attribute of concern. $\Diamond$

There are many quality attributes important to Web service operations. Table 2 organizes them into three categories: *runtime, business,* and *security* [70].

***Operation Interface***. Operation interface *OperationInterface* specifies a particular Web service capability through its name, purpose, domain of interest, locale of interest, quality attributes of concern, condition, signature and binding. Web services that implement the capability need to comply with what the interface has specified. The signature includes zero or one $Message_{in}$ and zero or one $Message_{out}$. $\Diamond$

The permutation of $Message_{in}$ and $Message_{out}$ determines the *mode* of the operation interface. There are four operation modes:

1. *one-way* - operation interface contains only $Message_{in}$
2. *notification* - operation interface contains only $Message_{out}$
3. *request-response* - operation receives $Message_{in}$ and then generates $Message_{out}$
4. *solicit-response* - operation generates $Message_{out}$ and then receives $Message_{in}$

**Table 2.1** QoWS Attributes

| Attribute Group | Attribute | Definition |
|---|---|---|
| Run-time | Response Time | Time between an invocation request is sent and processing results are returned |
| | Reliability | $N_{success}(op)/N_{invoked}(op)$ where $N_{success}$ is the number of times that $op$ has been successfully executed and $N_{invoked}$ is the total number of invocations |
| | Availability | UpTime($op$)/TotalTime where UpTime is the time $op$ was accessible during the total measurement time TotalTime |
| Business | Cost | Dollar amount to execute the operation |
| | Reputation | $\sum_{u=1}^{n} Ranking_u(op)/n$, $1 \leq$ Reputation $\leq 10$ where $Ranking_u$ is the ranking by user $u$ and n is the number of times $op$ has been ranked |
| | Regulatory | Compliance with government regulations, $1 \leq$ Regulatory $\leq 10$ |
| Security | Encryption | A boolean equal to *true* iff messages are encrypted |
| | Authentication | A boolean equal to *true* iff consumers are authenticated |
| | Non-repudiation | A boolean equal to *true* iff participants cannot deny requesting or delivering the service |
| | Confidentiality | List of parameters that are not divulged to external parties |

***Operation***. An operation embodies a specific implementation of an *OperationInterface* by a Web service. An operation can also make known of its need to invoke operations that implement an *OperationInterface*. In addition, an operation exhibits values for the quality attributes specified by the *OperationInterface*. ◇

The separation of operation from *OperationInterface* allows the same *OperationInterface* representing a shared capability to be implemented by multiple Web service operations.

***Web Service***. A Web service is defined by a tuple *(Name, Description, Operations, NFPs)*, where:

- *Name* is the name of the Web service;
- *Description* is a text summary about the service capabilities;
- *Operations* are a set of capabilities provided by the Web service; and
- *NFPs* stands for any non functional properties that help describe the Web service.

In addition, the Web service specifies values for relevant locale of interest attributes. ◇

In a Web service modeling a biological process, examples of non functional property include the declaration of the type of an entity that can provide the corresponding service and the description of the source information that the Web service model originates from.

The majority of constructs captured in Figure 2.1 are also included in OWL-S and WSMO. A major difference between ours and these standards is our proposed use of *OperationInterface*, which allows Web services to plug into one another declaratively at the operation level. Since *domain* is a grouping of functionalities, it is essentially a collection of corresponding *OperationInterfaces*.

# Chapter 3
# Hints from A Molecular Analogy

The first challenge faced in Web service mining is the problem of combinatorial explosion, which seems inevitable without specific queries to aid in the search process. Nature, however, has already provided us with ample examples on how it solves the problem of composition from bottom up. Molecular composition is in many ways similar to Web service composition. Like a composite service composed of component services, a molecule is formed from constituent atoms and/or simpler molecules. Certain molecules in the natural world are formed, because the reaction among constituent components under the surrounding condition happens to be more favorable for them to arise. The establishment of an analogy between the chemical world and the Web service world can yield important clues for efficient mining strategies.

In this chapter, we start with a simple analogy between the water molecule and a travel service in Section 3.1. We extend this analogy to a more complex situation in Section 3.2. In Section 3.3, we derive, from the phenomenon of molecular recognition, the concept of service/operation recognition, which later forms the foundation of our service screening algorithms. In Sections 3.4 and 3.5, we explore similarities of usefulness, side effects and effects of external conditions in both the chemical and Web service worlds. We explore further in Section 3.7 the similarity with respect to a process perspective between the two worlds. The identification of these similarities helps us establish a more holistic view when we later investigate and evaluate the various aspects of a service composition and service network identified out of our mining process. In Section 3.8, we look for strategies used in the drug discovery process that are applicable to Web service mining. Inspired by many of the similarities between the molecular and service worlds, we decided to apply our service mining framework back to the field of bioinformatics and, more specifically, the discovery of pathways linking biological processes. We describe a use case in Section 3.9 on how various biological processes can be modeled as Web services to pave the way for such application.

G. Zheng and A. Bouguettaya, *Web Service Mining: Application to Discoveries of Biological Pathways*, DOI 10.1007/978-1-4419-6539-4_3,
© Springer Science+Business Media, LLC 2010

## 3.1 A Simple Analogy

Web services share similarities in many ways with atoms and molecules in the natural world. At the most basic level, Web services are analogous to atoms, as illustrated in Figures 3.1(a) and (b). In the chemical world, oxygen and hydrogen atoms (Figure 3.1(a)) are two of the most basic building blocks in nature. An oxygen atom has six electrons in its outer shell and needs two more electrons to fill the shell. Meanwhile, a hydrogen atom has one electron in its outer shell. This electron can be shared with many other atoms including the oxygen atom. Under the right condition, the *supply-demand* relationship between the two types of atoms drives them to form *bonds* and, ultimately, a water molecule. Each of the two hydrogen atoms binds with the oxygen atom by sharing its sole electron with the oxygen. In return, the oxygen atom shares one of its six electrons with each of the two hydrogen atoms. This form of bonding between oxygen and hydrogen is called *covalent* bonding and is what holds a water molecule together.

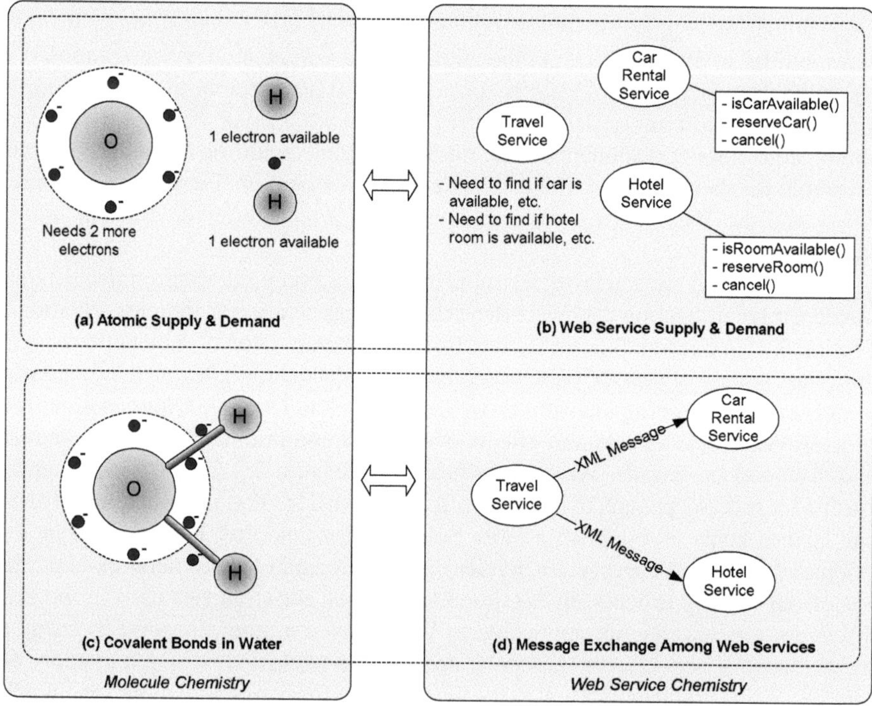


**Fig. 3.1** Molecules vs. Web Services

Similar supply-demand relationship is also what drives Web services together. Like a molecule that is composed of atoms or simpler molecules, a composite Web service is composed of simpler component Web services. Figure 3.1(b) shows a Web service example containing a *travel service*, a *car rental service* and a *hotel service*. The travel service needs to know, for example, whether a rental car is available within a given time period in a certain location. In addition, it also needs to know whether there is a room vacancy at the same proximity. These inquiries can be fulfilled by invoking operations provided by both the car rental and hotel services. Operation invocations are realized through the exchange of XML messages. Similar to electrons in the natural world, these messages bind component Web services into composite services. If we imagine each message as an electron, then this type of message exchange can help establish bonds between Web services. Here, we use bond as a notional concept to indicate the composability between two Web services.

## 3.2 A More Complex Analogy

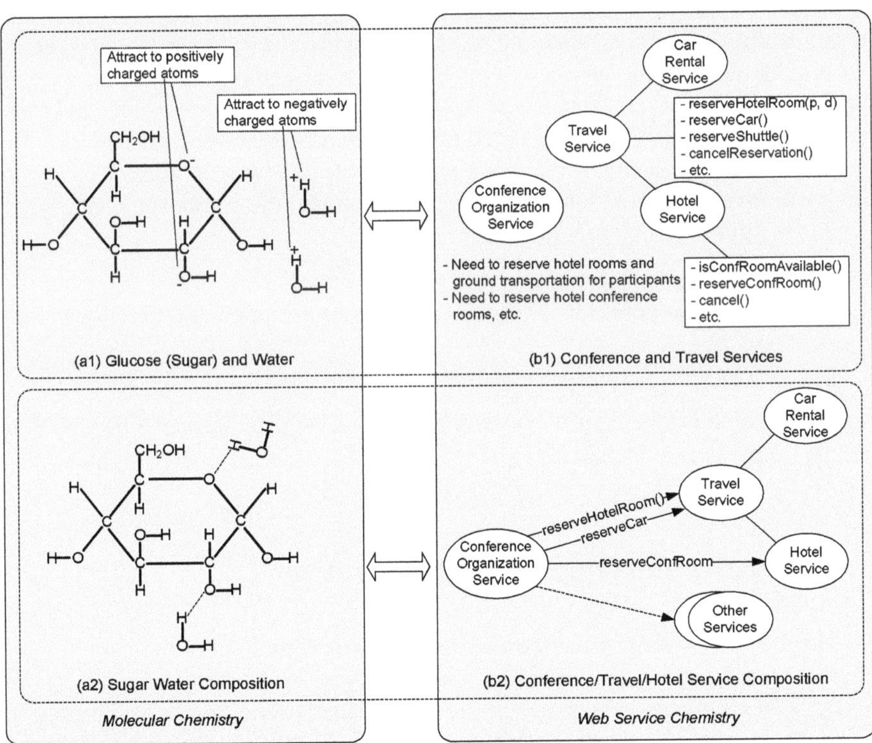


**Fig. 3.2** A More Complex Analogy

Similarity between the chemical and Web service worlds continues at a more complex level. For example, glucose (Figure 3.2(a1)) is a common sugar molecule present in many fruits and animal blood [83]. Its molecule structure contains a ring with five OH (called *hydroxyl*) groups. The oxygen atoms in the ring and the five hydroxyl groups are negatively charged. In the meantime, each of the two hydrogen atoms in every water molecule is positively charged. When these oppositely charged atoms are put together, they establish a new type of supply-demand relationship called *electrostatic attraction*. As a result of this attraction, several hydrogen bonds (dashed lines in Figure 3.2(a2)) are formed between the glucose molecule and water molecules. This mechanism allows glucose to dissolve easily in water. The composition of glucose and water results in some new properties. For example, the boiling temperature is raised while the freezing temperature is lowered. The lowering of the freezing temperature allows sugar to be used as an ice melter in the winter.[1] More importantly, the composition resulting from glucose or other vital nutrients dissolving in water allows such nutrients to be transferred through water to various parts of our body.

Likewise, the supply-demand relationship among Web services allows them to be composed at a more complex level. Figure 3.2(b1) shows that a *conference Web service* (CWS) needs to reserve conference rooms and help participants reserve hotel rooms and cars. In the meantime, the hotel and travel services established earlier can provide constituent services through their corresponding operations. As a result, XML messages are exchanged between the CWS and both the hotel and travel services (Figure 3.2(b2)), establishing bonds between the CWS and the other two services. Similar to chemical composition that leads to new properties, Web service composition leads to new functionalities. For example, the composite CWS is now capable of streamlining the following:

1. Collect conference reservations from participants;
2. Provide one-stop point for arranging ground transportation and lodging at a discount rate; and
3. Help synchronize participants' activities with the conference schedule.

The value added by the CWS is something that neither the travel nor the hotel service is able to provide alone.

## 3.3 Molecular and Operational Recognition

The bond establishment in neither world is arbitrary. Similar to the case where a chemical bond cannot be formed between two positively charged atoms, syntactically and/or semantically incompatible service operations cannot exchange messages. The analogy between the two worlds can be further explored in more complicated situations. In the natural chemical world, a phenomenon called *molecular*

---

[1] However, salt is more commonly used since it lowers the water's freezing temperature more effectively and is relatively more available and inexpensive.

*recognition* has been discovered [25]. Molecular recognition happens when two molecules recognize each other and interact to form bonds in between. Molecular recognition mechanism is preprogrammed and enabled in Deoxyribonucleic Acid (DNA) molecules in living organisms. It explains why a molecule (e.g. a protein molecule or *receptor*) is capable of recognizing and interacting with another molecule (e.g. a drug molecule or *ligand*), thus plays a crucial role in pharmaceutical drug discovery. For molecular recognition to happen, the following criteria must be met [73, 25]:

- *steric complementarity*, which requires that molecules of interest have compatible size and juxtaposed shape. This is like being able to insert a key into a lock, where the key is the ligand and the lock is the receptor. To be able to open the lock, we also need the following criteria,
- *electrostatic complementarity*. This requires that molecules of interest contain correctly placed and complementarily charged atoms.

In the pharmaceutical industry, molecular recognition is more specifically termed *pharmacophore*. Pharmacophore allows ligands to bind, interact, and modulate the activity of abnormal biological receptors that are the cause of undesirable medical conditions. To achieve pharmacophore, the following interaction is also needed:

- *hydrophobic interaction*, which takes place between ligands and receptors in our body, containing largely of water. The hydrophobic portions of the ligand are attracted to the hydrophobic pockets contained in the receptor. The arrangement of hydrophobic surfaces of the receptor allows it to limit the binding of inappropriate targets.

Similarly, Web services and their operations can recognize each other through both syntax and semantics. In the following, we introduce four types of recognition between Web services and operations.

**Direct Recognition.** A direct recognition is established between operations $op_a$ and $op_b$, if $op_a$ consumes an operation interface $op_{intf}$, which is implemented by $op_b$, as shown in Figure 3.3. In addition, $op_a$ and $op_b$ must be *mode*, *binding* and *message* composable [71]. ◇

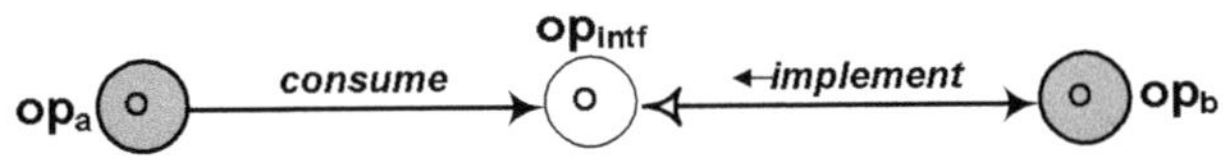


**Fig. 3.3** Direct Recognition between Web Service Operations

Mode composability states that the following pairs are composable: *notification* with *one-way* and *solicit-response* with *request-response*. Binding composability states that both operations should share the same protocol. Mode and binding composabilities are like being able to insert a key to a lock and are analogous to steric complementarity. Message composability states that the number of parameters and type of each parameter should match between the two operations. Message composability is like being able to open the lock with the correct key and is analogous to electrostatic complementarity and hydrophobic interaction.

In Figure 3.4, we illustrate three other service operation recognition mechanisms, namely *indirect recognition, promotion* and *inhibition*.

**Indirect Recognition**. A *target* operation (e.g., $op_{c2}$) indirectly recognizes a *source* operation (e.g., $op_{a1}$), if the source operation generates some or all input parameters (e.g., $p : T_a$ and $p : T_b$) of the target operation. $\diamondsuit$

We use the term *indirect* to indicate the fact that there is a potential need to relay parts of the *output* message from the source operation to parts of the *input* message to the target operation at the composition level.

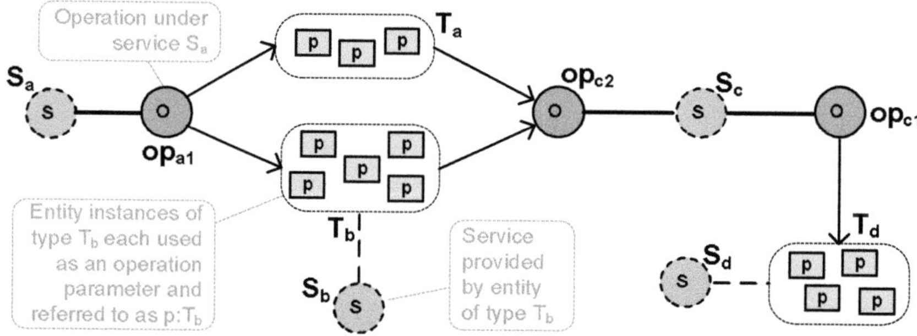


**Fig. 3.4**  Indirect Recognition, Promotion and Inibition between Web Service Operations

In the following two patterns, we assume that the availability of a service is directly proportionate to the quantity of entity instances that provide the service. Thus, the increase in such quantity tends to promote the service and reduction in such quantity tends to inhibit the service.

**Promotion**. When an operation (e.g., $op_{c1}$ of $S_c$) produces a substance (e.g., $p : T_d$ as output parameter of $op_{c1}$) that in turn provides a service (e.g., $S_d$), we say that the operation promotes the service. $\diamondsuit$

**Inhibition. Inhibition**. When an operation (e.g., $op_{c2}$ of $S_c$) consumes a substance (e.g., $p : T_b$ as input parameter of $op_{c2}$) that in turn provides a service ($S_b$), we say that the operation inhibits the service. $\diamondsuit$

**Operation Bond**. A *bond* is established between $op_s$ and $op_t$ for each input parameter $op_t$ can receive from $op_s$. We denote the set of bonds between $op_s$ and $op_t$ as $B(op_s \rightarrow op_t)$. If we refer to the set of all operations that $op_t$ recognizes as $OP_s(\rightarrow op_t)$, then

$$OP_s(\rightarrow op_t) = \{op \mid B(op \rightarrow op_t) \neq \phi\} \tag{3.1}$$

$op_s$ is said to be *completely bound* if it receives all its parameters from its bonds with other operation implementations. Otherwise, it is either *unbound* if none of its input parameters is received, or *partially bound* if only some of its parameters are received. ◇

Figure 3.5 shows that with indirect recognition, both operations can be of type *request-response* since the composed service can interact with both component services involved in the composition and relay parts of $Message_{out}$ from one component service to parts of $Message_{in}$ to the other. In some cases, a generated parameter might be synonymous to an input parameter. This happens when the two ontology nodes referred to by the generated parameter and the input parameter are synonymous. In such a case, a parameter conversion may be needed. In addition to relaying the message parameters, the composed service may need to coordinate or choreograph the invocation of component services to ensure the semantic and temporal correlations of these parameters. Indirect recognition between operations is analogous to electrostatic complementarity.

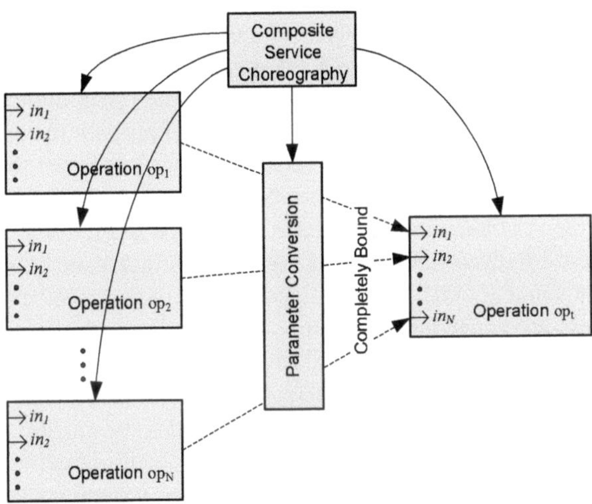


**Fig. 3.5** Indirect Recognition

Table 3.1 shows how the aggregate QoWS can be computed for multiple operations involved in the bonding of an operation. These are used later when determining which combination of operations exhibits the highest QoWS.

**Table 3.1**  Aggregate QoWS

| *QoWS attribute* | *Aggregation function* |
|---|---|
| Response time | $Max_{i=1}^{N} t(op_i), \sum_{i=1}^{N} t(op_i)$ |
| Availability | $\prod_{i=1}^{N} av(op_i)$ |
| Reliability | $\prod_{i=1}^{N} rel(op_i)$ |
| Cost | $\sum_{i=1}^{N} cost(op_i)$ |
| Reputation | $\frac{1}{N} \sum_{i=1}^{N} rep(op_i)$ |
| Regulatory | $\frac{1}{N} \sum_{i=1}^{N} reg(op_i)$ |
| Encryption | $\prod_{i=1}^{N} enc(op_i)$ |
| Authentication | $\prod_{i=1}^{N} aut(op_i)$ |
| Non-repudiation | $\prod_{i=1}^{N} nrep(op_i)$ |
| Confidentiality | $\prod_{i=1}^{N} con(op_i)$ |

## 3.4 Usefulness and Side Effects

*Usefulness* indicates how useful a chemical compound or a service composition is. It is what drives our desire to seek for compositions. The meaning of usefulness is sometimes *relative* and *subjective*. A chemical compound may be useful in one situation (e.g., water for drinking) but harmful in some other situations (e.g., water in a flood). Similarly, a broadcast service composed of registration service and email service may be considered useful to people who want to market their product, but annoying to those who don't want to receive spam emails.

Usefulness often comes with side effects as a by-product. For example, the drug discovery process involves finding lead compounds that may be only marginally useful (i.e., have some activity against a disease) and may have severe side effects. These side effects are caused by the compounds' ability to bind with body protein or nucleotide at unwanted points. Similarly, Web services composition may also introduce unwanted side effects. Composing e-commerce type of Web services may help streamline activities that would previously need to be carried out manually, it may also inadvertently compromise consumer's privacy if component Web services impose different standards for protecting consumers' identities. While being able to provide travel services, its trustworthiness as a whole may become questionable. A composition of biological Web services involving the service of a drug candidate may indicate a cure or a treatment for a certain disease or undesired condition, it may also introduce some unwanted side effects that potentially out-weight the benefits it brings.

## 3.5 Effects of External Conditions

External conditions such as light, temperature and pressure play an essential role in the physical states of materials and their chemical reactions. For water, temperature and pressure determine what physical state (i.e., vapor, liquid or ice) it will be in. For photosynthesis, light provides a necessary condition for chlorophyll in plant cells to convert water and carbon dioxide into sugar and oxygen.

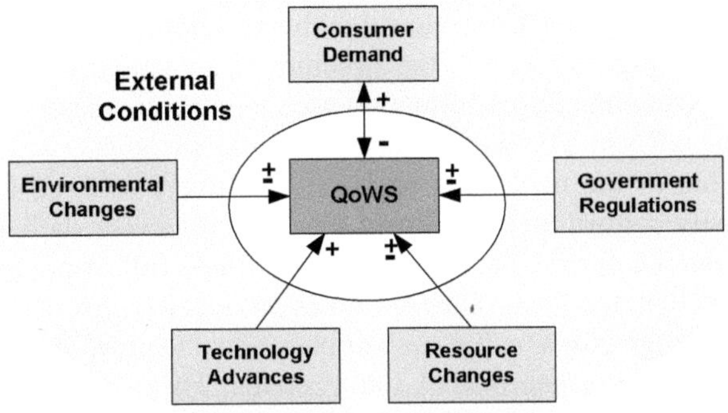


**Fig. 3.6** Effects of External Conditions

External conditions also play an important role in Web service formation. External conditions may include consumer demand, environment changes, technology advances, resource changes, and government regulations, etc., as illustrated in Figure 3.6. Consumer demand for certain Web services can fluctuate or shift. This may be due to the QoWS of existing services, availability of new services, and seasonal and environmental changes. QoWS of existing services has a positive effect on consumer demand, i.e., the better the QoWS, the higher the demand, though this relationship may saturate at some point. Availability of new services could have a negative effect on consumer demand for the existing services, i.e., the more choices people have, the less likely they are to opt to continue using the same service. Seasonal changes have a periodic effect on consumer demand. For example, every year more people choose to travel in the summer than winter. In return, consumer demand may have a negative effect on QoWS, i.e., the more demand a Web service receives from its consumers, the more likely the QoWS tends to degrade. For example, the price for summer flights tends to go up due to the scarcity of availability. Environmental changes may include natural disasters such as storms, flood and earthquakes, etc., all of which may result in temporary or long term regional shutdown of certain services such as airlines and railways. Technology advances tend to help improve QoWS and enhance people's experience with Web services. Resource changes include changes in labor supply, gasoline supply, etc. Labor disputes and global oil crisis can both have negative impact on service availability and cost. Government regulations may, for example, restrict pollution to a certain level. This may increase the operating cost of certain services. Shortly put, QoWS interplays with external conditions and can be affected by these conditions. As a result, QoWS is a function of external conditions.

## 3.6 A Eukaryotic Cell

In this section, we take a brief look at a *eukaryotic* cell, the basic building block of living organisms such as animals, plants and fungi, which have multiple cells [2]. We will refer to terminologies introduced in this example throughout the rest of the dissertation. As shown in Fig. 3.7, the structure of a eukaryotic cell has a *nucleus* that hosts the DNA containing information used to generate messenger Ribonucleic Acid (RNA), or mRNA. The nucleus also hosts the *nucleolus*, which produces *ribosomes*. Outside of the nucleus is an area collectively referred to as *cytoplasm* containing mostly *cytosol* and the rest of the cell organelles such as *endoplasmic reticulum* and *mitochondria*. The *cytosol* is the "soup" full of proteins responsible for controlling cell metabolism. The ribosomes produced by the *nucleolus* move out of the nucleus through pores on the *nuclear membrane* to positions on the *endoplasmic reticulum*. They translate mRNA into proteins. These proteins are then transferred to the *Golgi apparatus* in *transport vesicles* where they are further processed and packaged into *peroxisomes*, *lysosomes*, or *secretory vesicles*. *Peroxisomes* are responsible for protecting the cell from its own production of toxic hydrogen peroxide. As an example, white blood cells produce hydrogen peroxide to kill bacteria. The oxidative enzymes in peroxisomes break down the hydrogen peroxide into water and oxygen. *Lysosomes* contain hydrolytic enzymes necessary for intracellular

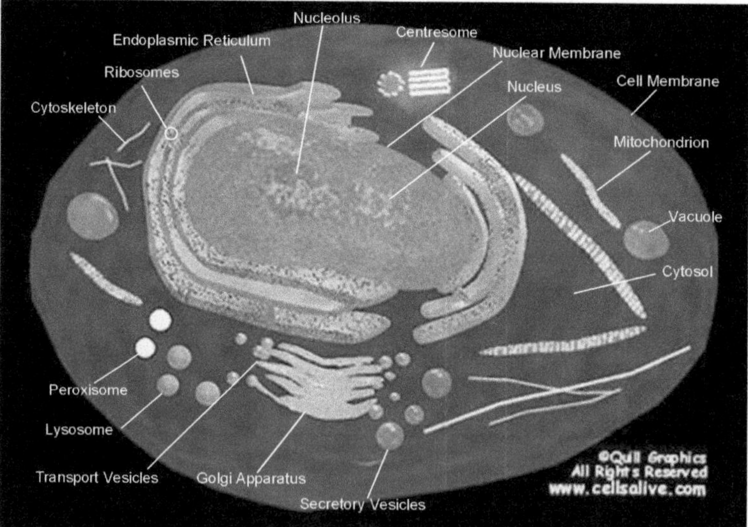


**Fig. 3.7**  Cell Structure and Organelles (adapted from *cellsalive.com*)

---

[2] The other type of living organisms such as bacteria are referred to as *prokaryotic*. They contain a single cell that does not have a nucleus.

digestion. In white blood cells that eat bacteria, lysosome contents are carefully released into the *vacuole* around bacteria and serve to kill and digest them. Cell secretions - e.g. hormones, neurotransmitters - are packaged in *secretory vesicles* at the *Golgi apparatus*. The *secretory vesicles* are then transported to the cell surface for release. *Centresome* is the "microtubule organizing center" within the cell and plays an essential role in cell division. *Cytoskeletons* are responsible for maintaining cell shape, locomotion and internal movement of cell organelles. *Mitochondria* are where food (sugar) is combined with oxygen to produce Adenosine Triphosphate (ATP), a high energy phosphate compound, which is the primary source of energy needed by the cell to move, divide and produce secretory products. Finally, the *cell membrane* enclosing the cell is where the *secretory vesicles* release cell secretions. It also contains ion channels responsible for the controlled entry and exit of ions. The cell beats when these ion channels open and close in an organized manner.

## 3.7  A Process Perspective

The analogy between the molecular and Web service worlds continues at a more complex process level as illustrated in Figure 3.8. In the chemical world, the DNA (Figure 3.8 (a1)) inside our cells (shown in Figure 3.7) provides a complete genetic blueprint that carries the information required to manufacture the enzyme proteins, which in turn are responsible for orchestrating our body's chemistry. The progression from DNA to protein involves a molecular assembly line [25] that follows a remarkable process. First, a gene-bearing portion of the DNA double helix is unraveled and information contained in one of the two single strands is transcribed into a *messenger RNA* (mRNA) through molecular recognition. The mRNA is then detached from the DNA strand and serves as a template in the protein building process (Figure 3.8 (a2)). This process involves a *transfer RNA*, or tRNA, which collects amino acids required to build the protein and carries them to the mRNA. Proteins are assembled on the mRNA one amino acid at a time. Links between adjacent amino acids called *peptide bonds* are established as the protein chain grows. The tRNA is then discharged and destroyed after its task is completed. Likewise, Web service composition can also involve a complex process. Figure 3.8 (b1) shows that a travel process flow template can be first constructed specifying the logic order of sequential, concurrent (i.e., AND Split *AND-S* and AND Join *AND-J*) and alternative (i.e., OR Split *OR-S* and OR Join *OR-J*) service invocations. Upon receiving a request, a process flow for a composite travel service is instantiated and grows (Figure 3.8 (b2)) based on this template. Here, the template is analogous to the blueprint carried by the mRNA and the process flow instance is analogous to the protein chain.

The similarities between Web services and *molecules* offer some interesting insights. They suggest that like molecules that compose from bottom up as if they are living beings, Web service can also be treated as living beings that *recognize* each other under the right conditions. The process analogy indicates that recognition-triggered service composition may extend to a flow network. Such a flow network

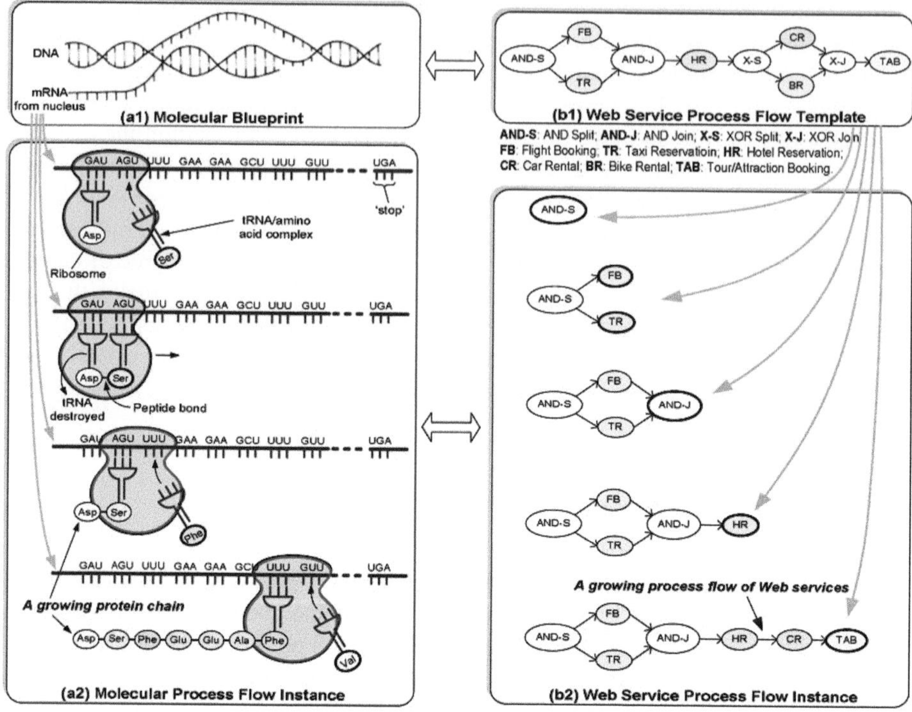



**Fig. 3.8**  Molecules and Web Service Dynamics ((a1)/(a2) based on [89] and [25])

can be either designed from top down or emerged from bottom up. As a result, instead of having to search for interesting and useful Web service compositions and composition networks exhaustively, there might be an easier way to just let these compositions and composition networks form "naturally" from bottom up, similar to what is happening in the natural world.

## 3.8 Drug Discovery Process

As we look for appropriate Web service mining techniques, the identification of many of the similarities between Web services and molecules has led us to the relevance of molecular composition methodologies and techniques resulted from the intense research effort invested in drug discovery. While many of the techniques are still evolving, the linkage between the two worlds has opened up new channels towards finding efficient techniques that might be suited for Web service mining. In this section, we look for suitable strategy from the drug discovery process. A typical drug discovery process [22] involves seven steps:

1. Disease selection,
2. Target hypothesis,
3. Lead compound identification and screening,
4. Lead optimization,
5. Pre-clinical trial,
6. Clinical trial, and
7. Pharmacogenomic optimization.

There are several interesting observations about the process described above. First, the drug discovery process has adopted the strategy of screening molecules (step 3) using "coarse-grained" filtering approach to quickly reduce the search space from the focused library of potential ligands to one that contains those most likely to bind to a protein target with high affinity. It then increases the computation complexity with better accuracy on a reduced search space for lead optimization (step 4). With a much smaller remaining space, the discovery process finally conducts more expensive clinical study for drug evaluation. This is a powerful strategy and can also apply well in the field of Web service mining.

As the number of Web services is expected to increase manifold, Web service mining will face the same combinational explosion challenge as has drug discovery. A similar Web service screening approach could take advantage of some "mining context" to scope down the searching space and identify potentially composable Web services in an early stage. The identification of the composability can be achieved using a "coarse-grained" ontology-based filtering mechanism. Automatic verification and objective analysis can be applied next in a reduced pool of candidate services. A more elaborate runtime simulation mechanism can then be applied towards composed Web service leads in a much smaller search space to investigate the relationships among various composition leads involved in the composition network. Finally, expensive subjective usefulness analysis involving human in the loop can be conducted in an even smaller search space to distinguish those that are truly useful.

## 3.9 Web Service Modeling of Biological Processes

The similarity between molecules and Web services motivated us to apply our mining framework back to the field of bioinformatics and, more specifically, the discovery of pathways linking biological processes. To achieve this, we first model these entities as Web services. We expect that these models will initially have sketchy details about known attributes and processes based on lab discoveries. As our knowledge increases and the modeling techniques continue to mature, the fidelity and completeness of these models can also be increased accordingly. One of the most challenging issues in modeling biological entities is how to approximate the richness of their processes and contextual uncertainties (e.g., varying temperature and fluidity of the surrounding environment) in a way that the models themselves would yield similar responses to the same stimuli or changes in the environment. Instead of trying to solve the issues of model accuracy and completeness, which by themselves are active research topics, we focus on how Web services are used to describe aspects of biological entities that we already know how to model.

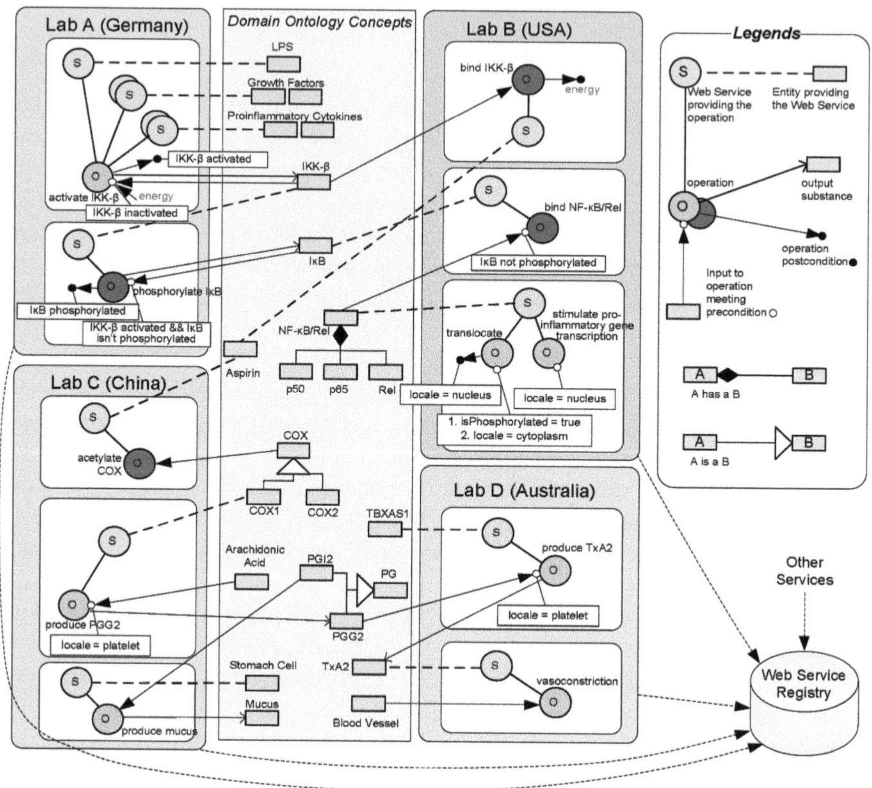


**Fig. 3.9**  Web Service Modeling

For illustration purposes, we use the scenario depicted in Figure 3.9 to show how different groups of Web service models can be produced and introduced independently to the Web service registry by different research laboratories and institutions from diverse geographic regions. The processes of various entities covered in this scenario are based on information obtained from [7], [23], [52] and [105]. Figure 3.9 shows that Labs A, B, C and D each carried out experiments focusing on studying the processes of specific set of biological entities. According to Lab A's experiment results, proinflammatory cytokines, LPS and growth factors can all activate IKK-$\beta$. Once activated, IKK-$\beta$ can in turn phosphorylate Inhibitory kappa B (I$\kappa$B). Based on such information, researchers from Lab A created several corresponding Web service models and published them into the Web service registry. Likewise, researchers from Lab B through their independent experiment, discovered that Aspirin can bind to IKK-$\beta$. This essentially blocks IKK-$\beta$'s normal processes. In addition, Lab B's results also revealed that I$\kappa$B can bind to Nuclear Factor kappa B (or NF-$\kappa$B/Rel), essentially blocking NF-$\kappa$B/Rel's normal processes, i.e., translocating from cytoplasm to the nucleus and then stimulating proinflammatory gene transcription in the cell nucleus. While the normal processes of IKK-$\beta$ would appear to be the missing link between the two findings and researchers from Lab B may be either unaware or unfamiliar with what exactly such processes might be, they were able to publicize their research observation nonetheless by creating Web service models for Aspirin, I$\kappa$B and NF-$\kappa$B/Rel and publishing them into the service registry. Similarly, researchers from both Labs C and D also created their Web service models based on their results and published them into the same service registry.

In Figure 3.9, processes of biological entities are modeled as operations of corresponding Web services. In general, an operation may take certain biological entities and an amount of energy from ATP as inputs. If the inputs and the service providing entity collectively meet the corresponding pre-conditions, then the processing of the operation becomes active. An example of a pre-condition would be that the activation of the *bind NF-$\kappa$B/Rel* operation of the I$\kappa$B Web service requires that I$\kappa$B itself be nonphosphorylated. The operation then takes $t$ units of time and $e_{in}$ units of energy to process $u_i$ units of each of its inputs $I$, where $i \in I$, generate $u_o$ units of each of its outputs $O$, where $o \in O$, and release $e_{out}$ units of energy. These outputs may be subjected to dissipation at a rate that is a function of the surrounding environment. At the end of the processing, some post-conditions may be effected triggering a change in the attributes of those entities that are involved after the operation finishes its processing [32]. Thus the net accumulation of each of the outputs will be the accumulation of the output from multiple invocations of the operation less what is being dissipated. To avoid ambiguities, domain ontology concepts are used to describe the type of inputs and outputs of these operations. Domain ontologies can capture relationships (e.g., inheritance and composition) among various concepts. For Web services covered in this scenario, such concepts include IKK-$\beta$, I$\kappa$B, NF-$\kappa$B/Rel, Cyclooxygenase (commonly referred to as COX), prostaglandin (PG), Aspirin, etc., as shown in the middle of Figure 3.9. Ontologies enable Web services and user applications to understand each other via communicating using an agreed-upon vocabulary. We assume that these ontologies are managed by domain experts using tools such as Ontolingua [44], Protégé [47], or Web Service Modeling Toolkit (WSMT) [12]. In addition, we assume that developers of biological Web services adhere to such ontologies when defining Web services.

# Chapter 4
# Web Service Mining Framework

Figure 4.1 shows our Web service mining framework, which can be figuratively described using the *"sow, grow, weed* and *harvest"* analogy.

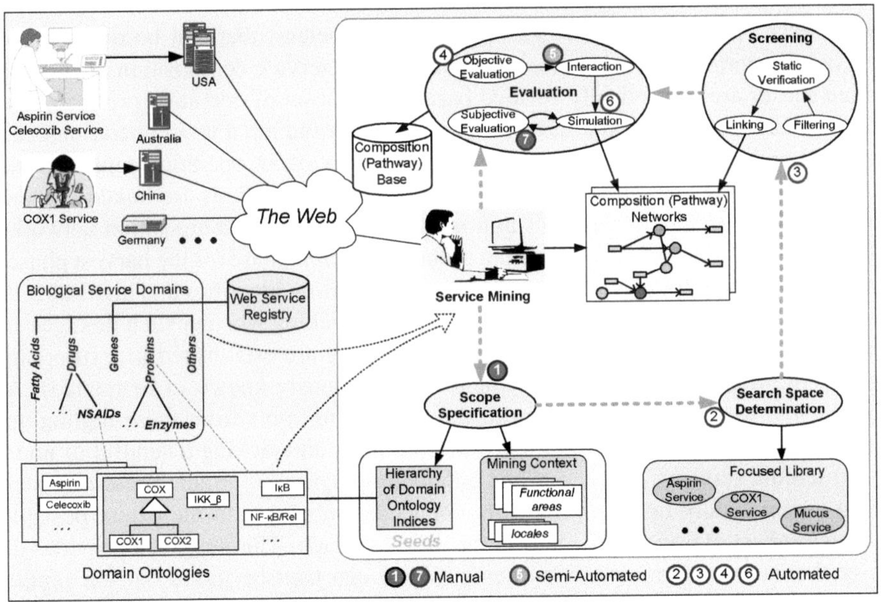



**Fig. 4.1** Service Mining Framework

The framework starts first with *scope specification*, a manual phase involving a domain expert defining the context of mining. We expect the domain expert to have a general idea about the "seeds" of 1. Web service functional areas (e.g., cell enzyme and drug functions) and 2. locales (e.g., heart, brain, etc. where these functions reside) that he/she is interested in mining. Such seeds are expected to grow into


G. Zheng and A. Bouguettaya, *Web Service Mining: Application to Discoveries of Biological Pathways*, DOI 10.1007/978-1-4419-6539-4_4,
© Springer Science+Business Media, LLC 2010

fruitful compositions (e.g., biological pathways) as the mining progresses. Within this phase, a hierarchy of *domain ontology indices* is established to speed up later phases in the mining process. Weights may be assigned to these seeds to differentiate user's interest in them and to help retain compositions grown out of those that the user is more interested in. Note that rather than specifying the exact goal of the compositions in pursuit as would a traditional Web service composition approach, scope specification does not limit what that goal should be. Consequently, any composition leads emerged within this scope will be pursued further.

Scope specification is followed by several automated phases. The first of these is *search space determination*. To help curb the problem of combinatorial explosion when faced with a large number of Web services, the mining context is used in this phase to identify a focused library of existing Web services as the initial pool for further mining. The next is the *screening* phase, which contains three sub-phases representing a *grow-weed-grow* cycle. In the *filtering* (first growing) sub-phase, Web services in the focused library would go through filtering algorithms for the purpose of identifying potentially interesting leads of service compositions or pathway segments. This is achieved through establishing linkages between Web services using a "coarse-grained" ontology-based filtering mechanism on only a subset of matching Web service characteristics (e.g., operation parameters) that can be quickly processed. In *static verification* (weeding) sub-phase, service composition leads identified earlier are semantically verified based on a subset of operation pre- and post-conditions involving binary variables (e.g., whether the input to an operation is activated) and enumerated properties (e.g., the locale of an operation input). In the *linking* (last growing) sub-phase, verified service compositions are linked together using link algorithms for establishing more comprehensive composition networks. These composition networks are input to the *evaluation* phase (or the harvest phase), which consists of four sub-phases. *Objective evaluation* identifies and highlights interesting segments of a composition network by checking whether such linkages are novel (i.e., previously unknown), and whether they are established in a surprising way (e.g., if the network links segments not previously known to be related). An *interactive* session follows next with the user taking hints from these highlighted interesting segments within a composition network and picking a handful of nodes representing services, operations and parameters to pursue further. These nodes are then automatically linked into a fully connected subgraph, to the extent possible, using a subset of nodes and edges in the original graph. This subgraph provides the user the basis to formulate hypotheses, which can then be tested out via *simulation*. In the case of pathway discovery, the simulation attempts to invoke relevant service operations, changing the quantity/attribute value of various entities involved in the composition network. Results from the simulation sub-phase are expected to reveal hidden relationships among the corresponding processes. These results are then presented to the user, whose *subjective evaluation* finally determines whether the subgraph, the basis of the user hypotheses, is actually useful. In some cases, the user may want to revise the simulation setting, re-execute the simulation and evaluate new simulation results. At the end, the user may want to introduce, in the case of pathway discovery, some of the discovered pathway subgraphs determined to be useful to a pathway base for future references. One use of such references may be in the area of building models for biological entities at a more complex level.

We present details of various phases of the mining process in the following three chapters. We combine scope specification and search space determination as two distinct steps in *pre-screening planning* and describe them in Section 4.1. We describe the details of the screening phase and the evaluation phase in Sections 4.2 and 4.3, respectively.


## 4.1  Pre-screening Planning

The mining process does not always have to be undertaken for *all* components within a global search space. In the pharmaceutical world, focused libraries have been used for a given receptor structure to reduce the search space when looking for leads. For Web service, the search space can be scoped down if we are only interested in finding composed services within certain functional areas. In addition, the search space can sometimes be scoped down further by locale of mining interest limiting the applicability of corresponding functions. We organize our pre-screening planning to contain two phases: *Scope Specification* and *Search Space Determination*.


### *4.1.1  Scope Specification*

The mining process starts with the *scope specification phase* where a composite Web service engineer optionally takes advantage of necessary subjective interestingness measures to bootstrap the mining process. The engineer may scope the mining activity by defining a list of functional areas and the locales where these functions reside. For example, the engineer may express a general interest in service compositions that involve travel, healthcare or insurance within the locale of the continental US. Another example may be to discover regulatory biological networks related to the locale of heart. Since different functional areas are drawn from corresponding domains, which may in turn rely on different ontologies, scope specification essentially determines a set of ontologies to use for the mining process. When presented with these ontologies, the engineer may choose to assign interestingness weights to various ontology nodes that he/she is interested in. As indicated in Figure 2.1, each domain also specifies what *OperationInterfaces* it contains. The engineer may optionally choose to assign interestingness weights to some of the *OperationInterfaces* that are of particular interest. The end product of scope specification is the *mining context* containing a list of relevant domains and locales limiting the applicability of corresponding functions within these domains. We formally define mining context $C$ in Eq. (4.1).

$$C = \{d(L) \mid d \in D\} \tag{4.1}$$

where

- $D$ is a set of Web service domains,
- $L$ is a set of locale attributes of mining interest,
- $d(L)$ is a domain carved out by $L$.

Consequently, if we use $\rightarrow$ to denote the relationship of *refers to*, then the set of all ontologies referred to in $C$ can be denoted as $Ont(C)$ and calculated using:

$$Ont(C) = \{ont \mid \exists d \in C \wedge d \rightarrow ont\} \tag{4.2}$$

and the set of all operation interfaces included in $C$ can be denoted as $OP_{intf}(C)$ and calculated using:

$$OP_{intf}(C) = \{op_{intf} \mid \exists d \in C \wedge op_{intf} \in d\} \tag{4.3}$$

We identify two scoping strategies.

- *Fixed Scope Mining*. The scope of mining identified at the beginning does not change during the mining process. This is used when the mining context is clearly defined and the search space can be easily determined.
- *Incremental Mining*. The initial scope of mining keeps expanding during the mining process. This is used when the mining context is unclear and cannot be easily specified at the beginning.


### 4.1.2 Search Space Determination

In both chemical and Web service worlds, mining scope determines the coverage of the search space when looking for composable components for the purpose of composition. Usually the more specific a scope is, the narrower a search space would be. Similar to the drug discovery process, the end product of our search space determination phase is a *focused library* consisting of Web services from service registry $R$ that are involved in mining context $C$. We formally define focused library $L$ in Eq. (4.4).

$$L = \{s \mid s \in R \wedge (s.operations \cap OP_{intf}(C) \neq \phi \ \vee$$
$$\exists op \in s.operations : op_{consume}(OP_{intf}) \cap OP_{intf}(C) \neq \phi)\} \tag{4.4}$$

where *s.operations* denotes the set of operations implemented by $s$ and $op_{consume}(OP_{intf})$ denotes the set of operation interfaces that are consumed by $op$. Thus Eq. (4.4) gives the focused library as the set of all Web services that either provide implementation(s) for some interface(s) in $OP_{intf}(C)$ or whose operation(s) consume(s) some implementation(s) of interface(s) in $OP_{intf}(C)$. The focused library thus covers the search space that is carved out based on the identified mining context. Algorithm 1 list the corresponding algorithm for obtaining the focused library.

---

**Algorithm 1** Focused Library Generation

---

**Input**: Mining context $C = \{d(L) \mid d \in D\}$, Web service registry $R$.
**Output**: Focused library $F$.
**Variable**: Context operations $OP(C)$.

1: **for all** $d(L) \in C$ **do**
2:     **for all** $op \in d(L)$ **do**
3:         add $op$ to $OP(C)$;
4:     **end for**
5: **end for**
6: **for all** $s \in R$ **do**
7:     **if** $s.operations \cap OP_{intf}(C) \neq \phi \ \vee \exists op \in s.operations : op_{consume}(OP_{intf}) \cap OP_{intf}(C) \neq \phi$ **then**
8:         add $s$ to $F$;
9:     **end if**
10: **end for**
11: return $F$;

---

## 4.2 Screening

Similar to drug discovery, Web service mining faces the challenge of combinatorial explosion due to the vast number of Web services that will be available on the Semantic Web. Inspired by the idea of lead compound screening during drug discovery (Section 3.8), we propose to use an ontology-based screening mechanism to address the same problem. Our approach uses a publish/subscribe mechanism to convert the traditional combinatorial screening problem into a spontaneous service and operation recognition problem. As a result, top down searches are transformed into bottom up matches. The screening phase in our service mining framework consists of three distinct sub-phases: *filtering*, *static verification* and *linking*. In this chapter, we present algorithms used in these three sub-phases.

### 4.2.1 Filtering

Algorithms used in the filtering sub-phase consist of two steps: *operation level filtering* and *parameter level filtering*.

### 4.2.1.1  Operation Level Filtering

Our operation level filtering mechanism is based on direct recognition. It uses operation interfaces within the mining context to serve as the medium for Web service operations to plug into each other. Figure 4.2 depicts the comparison between a naive exhaustive search mechanism and our operation level filtering mechanism. We compare the performance of these two mechanisms in section 4.2.1.4.

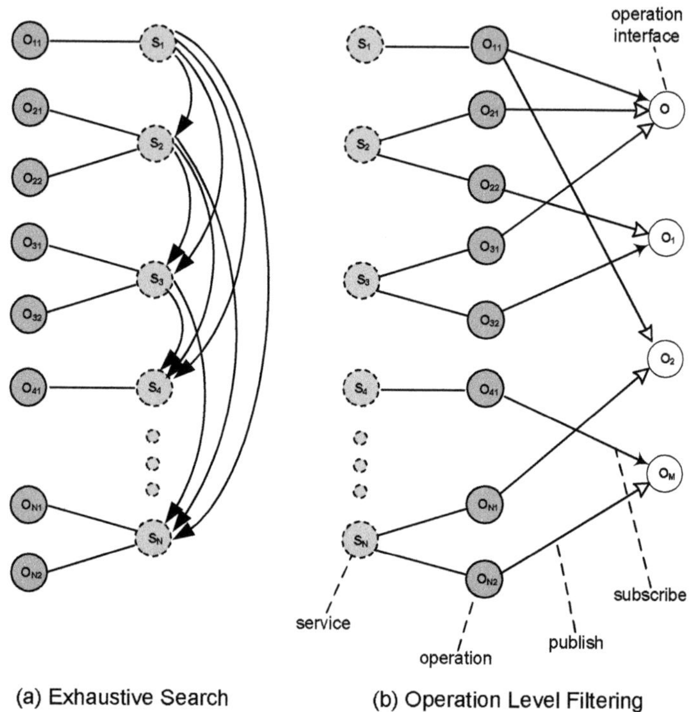


**Fig. 4.2**   Exhaustive Search vs. Our Operation Level Filtering

We list our operation level filtering algorithm in Algorithm 2. According this algorithm, Service operations that implement a particular interface publish their implementation through that interface (lines 6 - 9). Service operations that consume the implementation of an interface subscribe to that interface (lines 10 - 15). In addition, the service itself would subscribe to the ontology node that defines the type of its service providing entity. An operation agent is created (lines 1 - 3) for each operation interface to keep track of references to it from various operations. Algorithms 3 and 4 list our operation agent's functions for publication and subscription.

When publishing an operation that implements an interface, function *publish(op)* of the corresponding agent checks whether there is any subscriber

---

**Algorithm 2** Operation Level Filtering

---

**Input**: Context operation interfaces $OP_{intf}(C)$, focused library $F$.
**Output**: Leads of composed Web services $L$.
**Variables**: Leads from publication and subscription $L_{ps}$, operation interfaces consumed by $op, op_{consume}(OP_{intf})$.

  1: **for all** $op_{intf} \in OP_{intf}(C)$ **do**
  2:     create $Agent(op_{intf})$;
  3: **end for**
  4: **for all** $s \in F$ **do**
  5:     **for all** $op \in s.operations$ **do**
  6:        **if** $\exists op_{intf} \in OP_{intf}(C)$: op  implements  $op_{intf}$ **then**
  7:           $L_{ps} \leftarrow Agent(op_{intf}).\text{publish}(op)$;
  8:           $L.add(L_{ps})$;
  9:        **end if**
10:        **for all** $op_{intf} \in op_{consume}(OP_{intf})$ **do**
11:           **if** $op_{intf} \in OP_{intf}(C)$ **then**
12:              $L_{ps} \leftarrow Agent(op_{intf}).\text{subscribe}(op)$;
13:              $L.add(L_{ps})$;
14:           **end if**
15:        **end for**
16:     **end for**
17:     $k \leftarrow type(s.providerEntity)$;
18:     **if** $k \in Ont(C)$ **then**
19:        **if** $\neg \exists Agent(k)$ **then**
20:           create $Agent(k)$;
21:        **end if**
22:        $Agent(k).\textbf{subscribe}(s)$;
23:     **end if**
24: **end for**

---

---

**Algorithm 3** Operation Agent Function for Publication

---

*publish(op)*
**Input**: Web service operation $op$ providing implementation for the operation interface that this agent represents.
**Output**: Leads of composed Web services $L_{ps}$.
**Variable**: A composed service $cs$.

  1: $publishers.add(op)$;
  2: **if** $subscribers \neq \phi$ **then**
  3:     **for all** $op' \in subscribers$ **do**
  4:        $cs \leftarrow generateLead(op', op)$;
  5:        $L_{ps}.add(cs)$;
  6:     **end for**
  7: **end if**
  8: return $L_{ps}$;

---

---

**Algorithm 4** Operation Agent Function for Subscription

---

*subscribe(op)*
**Input**: Web service operation *op* interested in invoking the operation interface that this agent represents.
**Output**: Leads of composed Web services $L_{ps}$.
**Variable**: A composed service *cs*

1: *subscribers.add(op)*;
2: **if** *publishers* $\neq \phi$ **then**
3:    **for all** $op' \in publishers$ **do**
4:       $cs \leftarrow generateLead(op, op')$;
5:       $L_{ps}.add(cs)$;
6:    **end for**
7: **end if**
8: return $L_{ps}$;

---

to the operation. If so, it tries to establish a service composition lead using direct operation recognition between the publisher and the subscriber. Similarly, when a service operation subscribes to an operation interface that it consumes, function *subscribe(op)* checks whether there is any publisher that implements the operation. If so, it tries to establish a composed service lead between the subscriber and the publisher.

### 4.2.1.2 Parameter Level Filtering

Figure 4.3 illustrates our parameter level filtering mechanism that targets three types of Web service and operation recognition: *promotion, inhibition* and *indirect recognition*, as described in Section 3.3. We consider three types of matching between parameters $p_1$ and $p_2$, whose data types refer to domain ontology index nodes $n_a$ and $n_b$, respectively:

- *Exact match or synonym*: $n_a = n_b$. One index node is created for all synonymous ontology nodes.
- *Is-a*: $n_a$ is a child of $n_b$
- *Has-a*: $n_a$ has a component $n_b$

We assume that the above relationships among parameter types are already declared in domain ontologies and consequently domain ontology indices. Therefore, these relationships can be automatically detected. Figure 4.4 gives examples of biological processes modeled as Web services and illustrates how we keep track of publications and subscriptions at the parameter level using an ontology extension. Several Web service models are shown for biological entities including Aspirin, COX1, Peroxidase, PGI2 and Stomach Cell. The types of these service providing entities can be drawn from domain ontologies such as those for drug, protein, fatty acids, etc. To speed up the processing of type checking, we establish indices to domain ontology concepts as shown in the shaded box in the middle of the figure. To

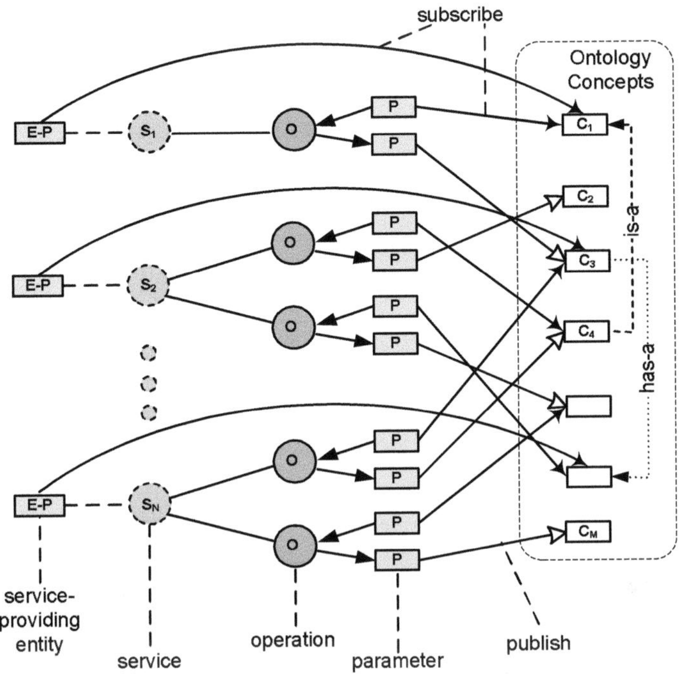



**Fig. 4.3**  Parameter Level Filtering

keep track of operation recognition that occurs at the parameter level, we extend
the capability of a conventional ontology by attaching a *node agent* to every ontology index node that has been referenced by at least one operation parameter or
service providing entity, as illustrated in the dashed box on the right of Figure 4.4.
To increase the hit rate and reduce computation complexity of our parameter level
filtering algorithm, we ignore operation pre- and post-conditions. We consider these
conditions during the static verification sub-phase in Section 4.2.2 and the simulation sub-phase in Section 5.4.0.9 when the pool of candidate is presumably much
smaller.

Algorithm 5 shows the parameter level filtering algorithm that uses node agents
to keep track of these references. Since ontology nodes are used to describe the type
of operation parameters, a parameter is considered an instance of such a node. When
an Web service operation is introduced in the mining process, each of the operation's
output parameters will publish to an ontology node it is an instance of (lines 3 to
11). For example, output parameters of COX1 service operation *producePGG2* in
Figure 4.4 will publish to node *PGG2*. Similarly, each of its input parameters will
subscribe to an ontology node it is an instance of (lines 12 to 20). For example,
the input parameter of COX1 service operation *producePGG2* in Figure 4.4 will
subscribe to node *Arachidonic Acid*.

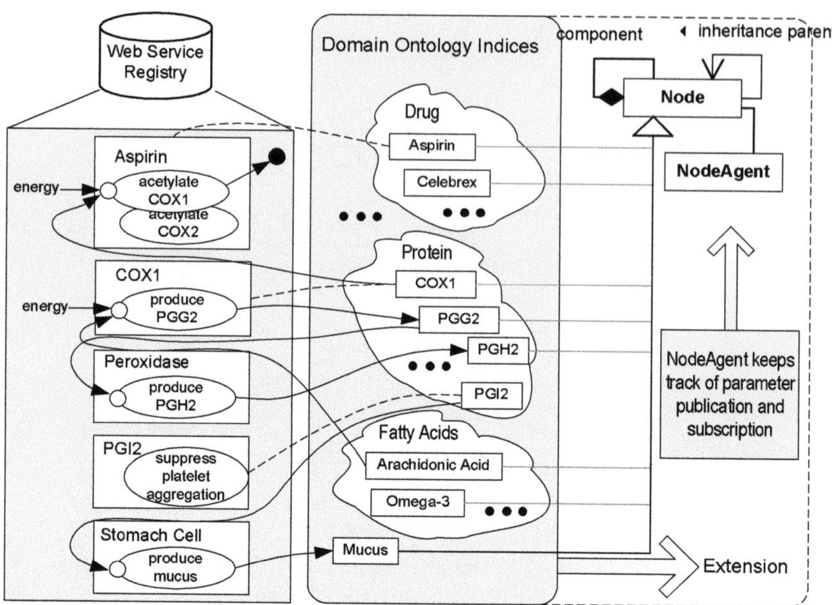


**Fig. 4.4** Ontology Extension for Web Service Screening

As an operation is introduced, its corresponding parameters trigger the necessary subscriptions and publications. Since each ontology node inherits the *Node* class, it is capable of keeping track of its own subscribing and publishing parameters. This tracking enables a subscribing parameter of the node to become essentially notified of the presence of a publishing parameter. As a result, the operation which the subscribing parameter belongs recognizes the operation which the publishing parameter belongs, and a bond is established. Algorithms 6, 7, and 8 list the functions provided by a node agent to register the publication of an output parameter, subscription of an input parameter, and subscription of a service an instance of the corresponding ontology node would provide.

Within an ontology, publication and subscription on a node can sometimes propagate to other nodes within the ontology network. This happens when the node is involved in an inheritance or compositional relationship with other nodes. For example, if operation A needs *COX* as one of its input parameters and consequently subscribes to node *COX* and operation B generates *COX1* and consequently publishes to node *COX1*, then the publication on node *COX1* also propagates up the inheritance tree so operation A becomes aware of operation B on node *COX*. In general, publication propagates down a composition tree (lines 2 to 4 in Algorithms 6) and up an inheritance tree (lines 5 to 7 in Algorithm 6), while subscription propagates up a composition tree (lines 2 to 4 in Algorithm 7) and down an inheritance tree (lines 5 to 7 in Algorithm 7) in the ontology.

---

**Algorithm 5** Parameter Level Filtering

---

**Input**: context ontologies $Ont(C)$, focused library $F$
**Variables**: ontology type $type(p)$ parameter $p$ refers to.

1:  **for all** $s \in F$ **do**
2:      **for all** $op \in s.operations$ **do**
3:          **for all** $p_{out} \in op.message_{out}$ **do**
4:              $k \leftarrow type(p_{out})$;
5:              **if** $k \in Ont(C)$ **then**
6:                  **if** $\neg \exists Agent(k)$ **then**
7:                      create $Agent(k)$;
8:                  **end if**
9:                  $Agent(k)$.**publish**($\mathbf{p_{out}}$);
10:                **end if**
11:          **end for**
12:          **for all** $p_{in} \in op.message_{in}$ **do**
13:              $k \leftarrow type(p_{in})$;
14:              **if** $k \in Ont(C)$ **then**
15:                  **if** $\neg \exists Agent(k)$ **then**
16:                      create $Agent(k)$;
17:                  **end if**
18:                  $Agent(k)$.**subscribe**($\mathbf{p_{in}}$);
19:              **end if**
20:          **end for**
21:      **end for**
22:      $k \leftarrow provider(s)$;
23:      **if** $\neg \exists Agent(k)$ **then**
24:          create $Agent(k)$;
25:      **end if**
26:      $Agent(k)$.**subscribe**($\mathbf{s}$);
27: **end for**

---

**Algorithm 6** Registering the Publication of an Output Parameter through a Node Agent

---

$publish(p_{out})$
**Input**: $p_{out}$ of data type that this agent represents.

1:  $publishers.add(p_{out})$;
2:  **for all** $n \in CompositionalChildren$ **do**
3:      $Agent(n).publish(p_{out})$;
4:  **end for**
5:  **if** $inheritanceParent \neq \phi$ **then**
6:      $Agent(inheritanceParent).publish(p_{out})$;
7:  **end if**
8:  **for all** $p_{in} \in parameterSubscribers$ **do**
9:      **if** $p_{in}.operation \neq p_{out}.operation \wedge p_{out} \notin p_{in}.bonds$ **then**
10:        $p_{in}.bonds.add(p_{out})$;
11:     **end if**
12: **end for**
13: **for all** $s \in serviceSubscribers$ **do**
14:     $s.promotors.add(p_{out})$;
15: **end for**

---

**Algorithm 7** Registering the Subscription of an Input Parameter through a Node Agent

---

$subscribe(p_{in})$
**Input:** $p_{in}$ of data type that this agent represents.
1:  $subscribers.add(p_{in})$;
2:  **for all** $n \in CompositionalParents$ **do**
3:     $Agent(n).subscribe(p_{in})$;
4:  **end for**
5:  **for all** $n \in inheritanceChildren$ **do**
6:     $Agent(n).subscribe(p_{in})$;
7:  **end for**
8:  **for all** $p_{out} \in parameterPublishers$ **do**
9:     **if** $p_{in}.operation \neq p_{out}.operation \wedge p_{out} \notin p_{in}.bonds$ **then**
10:       $p_{in}.bonds.add(p_{out})$;
11:     **end if**
12: **end for**
13: **for all** $s \in serviceSubscribers$ **do**
14:     $s.inhibitors.add(p_{in})$;
15: **end for**

---

**Algorithm 8** Node Agent Function for Registering the Subscription of a Service

---

$subscribe(s)$
**Input:** $s$ by entity of data type that this agent represents.
1:  $serviceSubscribers.add(s)$;
2:  **for all** $p \in publishers$ **do**
3:     $s.promotors.add(p)$;
4:  **end for**
5:  **for all** $p \in subscribers$ **do**
6:     $s.inhibitors.add(p)$;
7:  **end for**

---

**Algorithm 9** Lead Generation

---

**Input:** Focused library $F$.
**Output:** Leads of composed Web services $L$.
**Variables:** Leads from publication and subscription $L_{ps}$,
1:  **for all** $s \in F$ **do**
2:     **if** $s.promoters \neq \phi$ or $s.inhibitors \neq \phi$ **then**
3:       $L_{ps} \leftarrow s.generateLeads()$;
4:       $L.add(L_{ps})$
5:     **end if**
6:     **for all** $op \in s.operations$ **do**
7:       **if** $op.isCompletelyBound()$ **then**
8:         $L_{ps} \leftarrow op.generateLeads())$;
9:         $L.add(L_{ps})$
10:       **end if**
11:     **end for**
12: **end for**
13: return $L$;

---

Algorithm 9 shows the algorithm used to generate leads of service compositions or pathway segments. Function *isCompletelyBound*() returns true if all input parameters of the operation is bound. When applied to pathway discovery, this function is degenerated into checking whether the operation is simply bound with any other operation. Function *generateLeads*() for a service generates a lead tree rooted at a service listing as its child nodes promoting operations outputting entity of its type or inhibiting operations consuming entity of its type. Function *generateLeads*() for an operation generates a lead tree rooted at an operation containing as its child nodes operations whose output parameters match its input parameters.

### 4.2.1.3 Filtering in Fixed Scope Mining and Incremental Mining

The filtering algorithms at both the operation and parameter levels can be applied after all operation interfaces for the mining context have been included in fixed scope mining or as more operation interfaces are identified and added to $OP_{intf}(C)$ due to the expansion of mining context in incremental mining. This allows incremental mining to be practically a more flexible approach than fixed scope mining. As the mining context settles down in incremental mining, the introduction of new operation interfaces to $OP_{intf}(C)$ will stop and expansion of the focused library will also cease. As a result, the filtering of Web services will eventually reach a closure.

### 4.2.1.4 Complexity Analysis

We compare the computation complexity of the screening algorithms against a naive exhaustive search algorithm using operation recognition at both the operation and parameter levels. Table 4.1 lists relevant variables used in our complexity analysis.

If we refer to the size of collection *s.operations* as $|S|$, then the time to carry out a hashtable-based check of the $\in$ operation (lines 6 and 11 in Algorithm 2) is $O[log(|S|)]$. In the following we first analyze the performance of the traditional exhaustive search mechanism (see Figure 4.2 (a)). An operation level composability check will iterate through all services in the scope and check each service against all other services. For a pair of services $s_1$ and $s_2$, it checks whether $s_1$'s operation could consume any of $s_2$'s operations, vice versa. There are two ways to match up operations from $s_1$ and $s_2$. The first is to iterate through $s_1$'s operations and for each operation iterate through $N_{oc}$ to see if an consumable operation is in $s_2$'s operation set. The second is to iterate through $s_1$'s operations and then $s_2$'s operations. For each operation found in $s_2$'s operation set, check whether it is consumable by the one found in $s_1$'s operation set. Thus, the time to perform operation level comparison using an exhaustive search is:

$$T_{of} = O[N_{ws}^2 \times min(N_{oi} \times N_{oc} \times logN_{oi}, N_{oi}^2 \times logN_{oc})]$$

**Table 4.1** Symbols and Parameters

| Variables | |
|---|---|
| $N_{opIf}$ | Number of operation interfaces in the mining context |
| $N_{pin}$ | Average number of input parameters to an operation |
| $N_{pout}$ | Average number of output parameters from an operation |
| $N_{ws}$ | Number of Web services in the focused library |
| $N_{oi}$ | Average number of operation interfaces each Web service implements |
| $N_{oc}$ | Average number of operation interfaces each operation consumes |
| $N_{ops}$ | Average number of operation subscribers to an operation interface |
| $N_{opp}$ | Average number of operation publishers to an operation interface |
| $N_{ns}$ | Average number of parameter subscribers to an ontology node |
| $N_{ss}$ | Average number of service subscribers to an ontology node |
| $N_{np}$ | Average number of publishers to an ontology node |
| $|Ont|$ | Size of domain ontologies |
| ***Performance measurement parameters*** | |
| $T_{of}$ | Time for operation filtering |
| $T_{mf}$ | Time for message/parameter filtering |
| $T$ | Total screening time ($T = T_{of} + T_{mf}$) |

We now analyze the performance of our operation level filtering algorithms. According to Algorithm 3, the time to perform $publish(op)$ is $O(N_{ops})$. Likewise, from Algorithm 4, the time to perform $subscribe(op)$ is $O(N_{opp})$. Thus, $T_{of}$ can be calculated according to Algorithm 2:

$$T_{of} = O[N_{opIf} + N_{ws} \times (N_{oi} \times (logN_{opIf} + N_{ops} + N_{oc} \times (logN_{opIf} + N_{opp})) + log(|ont|))]$$
$$= O[N_{ws} \times (N_{oi} \times (N_{ops} + N_{oc} \times N_{opp} + (1 + N_{oc}) \times logN_{opIf}) + N_{opIf} + log(|ont|))]$$

Comparing the performance of our filtering algorithms with that of an exhaustive search algorithm, we see that when $N_{opIf}$ is relatively small and stable as compared to $N_{ws}$, $T_{of}$ in our filtering algorithm is linear to $N_{ws}$, while $T_{of}$ in a traditional exhaustive search is exponential to $N_{ws}$. At the parameter level, the time $T_{mf}$ for the exhaustive algorithm can be calculated using:

$$T_{mf} = O[N_{ws}^2 \times N_{oi}^2 \times min(N_{pin} \times logN_{pout}, N_{pout} \times logN_{pin})]$$

The time $T_{mf}$ for our filtering algorithm according to Algorithms 5, 6, 7 and 8 is:

$$T_{mf} = O[N_{ws} \times (N_{oi} \times (N_{pout} \times (N_{ns} + N_{ss}) + N_{pin} \times (N_{np} + N_{ss})) + N_{np} + N_{ns})]$$

Similar to operation level filtering, $T_{mf}$ from our parameter level filtering algorithms is linear to $N_{ws}$, while $T_{mf}$ in the traditional exhaustive search is exponential to $N_{ws}$. We list $T_{of}$ and $T_{mf}$ for both our algorithms and the traditional exhaustive search in Table 4.2.

**Table 4.2** Performance Comparison

| *Our Screening Algorithm* |
| --- |
| $T_{of} = O[N_{ws} \times (N_{oi} \times (N_{ops} + N_{oc} \times N_{opp} + (1 + N_{oc}) \times logN_{oplf}) + N_{oplf} + log(\|ont\|))]$ |
| $T_{mf} = O[N_{ws} \times (N_{oi} \times (N_{pout} \times (N_{ns} + N_{ss}) + N_{pin} \times (N_{np} + N_{ss})) + N_{np} + N_{ns})]$ |
| *Exhaustive Search* |
| $T_{of} = O[N_{ws}^2 \times min(N_{oi} \times N_{oc} \times logN_{oi}, N_{oi}^2 \times logN_{oc})]$ |
| $T_{mf} = O[N_{ws}^2 \times N_{oi}^2 \times min(N_{pin} \times logN_{pout}, N_{pout} \times logN_{pin})]$ |

#### 4.2.1.5 Complexity Simulation

We conducted experiment to simulate the performance of the filtering algorithms. We used an XP machine with duo core 2.8GHz. We focus in our experiment on investigating the relationship between the total processing time of the filtering algorithms and the number of Web services that are used as inputs to these algorithms. Table 4.3 lists the configuration variables used in our experiment.

**Table 4.3** Experiment Settings for Filtering Algorithms Performance Simulation

| Variable | Value or Range |
| --- | --- |
| Number of domains | 50 |
| Number of operation interfaces per domain | 100 |
| Number of services $n_s$ | $100 \times 2^{iteration\_number-1}$ |
| Domain ontology index nodes to services ratio $s\_num$ | num ranges from 1/3 to 3 |
| Number of operations per service | $1 - 5$ uniformly distributed |
| Number of operation interfaces consumed by operation | $0 - 2$ uniformly distributed |
| Input parameters per operation | $1 - 10$ uniformly distributed |
| Output parameters per operation | $1 - 10$ uniformly distributed |

We use $n_s$ to denote the number of services in the input and $s\_num$ the ratio of ontology index nodes to services. For each $s\_num$, we iterate through $n_s$, which starts at 100 and doubles its values for each subsequent iteration, as indicated in Table 4.3. For each pair of $(n_s, s\_num)$, we run through the filtering algorithms 10 times. We then take the averages of the total processing time from these runs and plot them in

Figure 4.5. From simulation results in both Figures 4.5, we see that the total filtering time is linear to the number of services used as input.

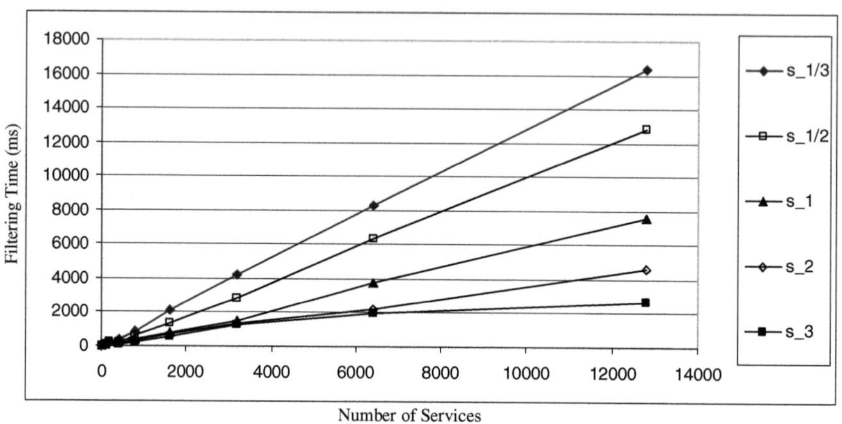



**Fig. 4.5**  Filtering Complexity Simulation: Processing Time vs. Number of Services

We note from Figure 4.5 that as the ratio of domain ontology index nodes to services decreases, the total filtering time tends to increase. This observation can be better expressed in Figure 4.6 and is consistent with our analytical model (see Tables 4.1 and 4.2). As the ratio of services to ontology index nodes increases, each index node is expected to have an increased number of parameter subscribers ($N_{ns}$), service subscribers ($N_{ss}$) and publishers ($N_{np}$). According to Eq. 4.5, $T_{mf}$ is linear to $N_{ns}$, $N_{ss}$ and $N_{np}$.

### 4.2.2 Static Verification

Various measures [71] have been proposed to determine whether two operations are composable at both syntactic and semantic levels. These measures can be used to determine whether a *direct recognition*-based composition is actually valid. For *promotion* and *inhibition*-based compositions, they are valid because the entities of interest provide the corresponding services by declaration. In this section, we focus on how the validity of an *indirect recognition*-based composition can be determined in the verification phase. We denote $comp(OP_s, op_t)$ as an operation composition involving a set of source operations $OP_s$ providing input parameters to target operation $op_t$, where $OP_s \subset OP_s(\rightarrow op_t)$. In order for $comp(OP_s, op_t)$ to be valid, the following must be true:

$$\forall op_s \in OP_s, \Gamma[op_s.L, op_t.L] \neq 0 \tag{4.5}$$

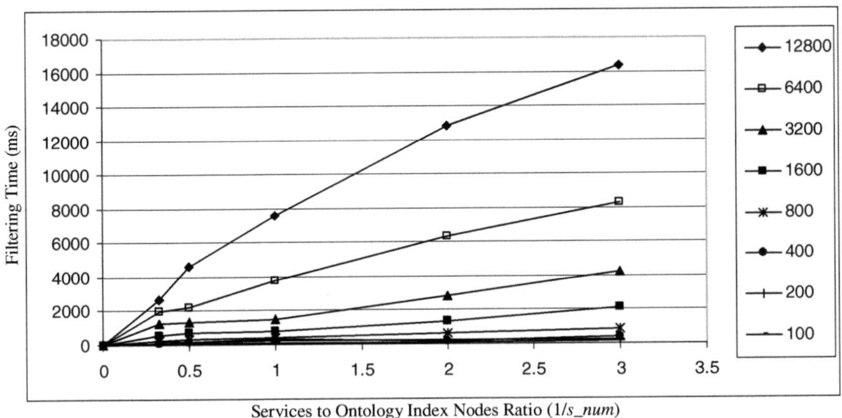

Services to Ontology Index Nodes Ratio (1/s_num)

**Fig. 4.6**  Filtering Complexity: Processing Time vs. Services to Ontology Index Nodes Ratio

where $\Gamma$ is a domain expert-determined correlation function that measures the relevancy (i.e., 1 for the same and non-zero for related) of two locales. Eq. (4.5) states that in order for the composition to be valid, each of the source operations must have a locale that correlates to that of the target operation. A relevant example in bioinformatics would be to make drug molecules effective (or compose with disease causing molecules) by sending them to where the disease cells are located.

If we use $B^t_{selected}(op_s \rightarrow op_t)$ to denote a set of $op_t$'s input parameters (i.e., target parameters) covered by a selected subset of $B(op_s \rightarrow op_t)$ for $comp(OP_s, op_t)$, $B^s_{selected}(op_s \rightarrow op_t)$ the corresponding output parameters from $op_s$, and $EA_{comp}$ external attributes involved in $comp(OP_s, op_t)$, then in order for the composition to be valid, the following must also be true:

$$\exists F(OP_s) : \forall B^t_{selected}(op_s \rightarrow op_t) \; where \; op_s \in OP_s,$$

$$\sum_{\substack{op_s \in OP_s \\ f \in F(OP_s)}} \{(C_{post}(f(op_s)) \; on \; B^s_{selected}(op_s \rightarrow op_t) \; and \; EA_{comp})\}$$

$$\supset \{C_{pre}(op_t) \; on \; B^t_{selected}(op_s \rightarrow op_t) \; and \; EA_{comp}\} \qquad (4.6)$$

$F(OP_s)$ refers to composition-specific mediations that need to be applied to $OP_s$ so that the combined post-conditions $C_{post}$ of $OP_s$ cover the space carved out by the pre-conditions $C_{Pre}$ of $op_t$ if all $B^s_{selected}(op_s \rightarrow op_t)$ are replaced by corresponding $B^t_{selected}(op_s \rightarrow op_t)$. Consequently, invocation of $op_t$ is activated.

The filtering algorithm, when applied to pathway discovery, can identify pathway segment leads, the building blocks for pathways. Each segment may be used multiple times later in building different pathways or different sections of the same pathway. To reduce the chance of contamination during pathway construction by invalid segments, we present algorithm to check for their validity in Algorithm 8.

---

**Algorithm 10** Static Verification

---

**Input**: Leads of compositions $L_{cs}$
**Output**: Composition network.
**Variables**: composition lead $l_{cs}$, root operation node $rootOp$, root service node $s$, entity $e$, parameter subNode $para$, and operation subNode $op$.

```
 1: for all l_cs ∈ L_cs do
 2:     if l_cs.rootNode is an operation then
 3:         rootOp ← l_cs.rootNode
 4:         for all parameter subNode para of rootOp do
 5:             for all provider operation subNode op of para do
 6:                 if ¬overlap(rootOp.getStaticPreConds(), op.getStaticPostConds()) then
 7:                     remove op as a subNode from para;
 8:                 end if
 9:             end for
10:         end for
11:     end if
12:     if l_cs.rootNode is a service then
13:         s ← l_cs.rootNode;
14:         e ← entity subNode of s;
15:         for all operation subNode op of e do
16:             if op is a promoter of s then
17:                 if ¬∃op' ∈ s.operations : overlap(op'.getStaticPreConds(), op.getStaticPostConds())
                    then
18:                     remove op as a promoter of s;
19:                 end if
20:             end if
21:             if op is an inhibitor of s then
22:                 if ¬∃op' ∈ s.operations : ¬overlap(op'.getStaticPreConds(), op.getStaticPostConds())
                    then
23:                     remove op as an inhibitor of s;
24:                 end if
25:             end if
26:         end for
27:     end if
28: end for
```

---

The verification algorithm relies on checking *static* conditions involving binary variables and enumerated properties. For each of the composition leads, we check to see if it starts with an operation (lines 02-11) or a service (lines 12-27). In the first case, we check whether the static post-conditions of an operation $op$, which is identified as one recognized by the root operation $rootOp$ of the lead, overlap with the static pre-conditions of $rootOp$. If not, then $op$ is removed from the list of operations that $rootOp$ recognizes. In the second case, we check for each of the operations identified as either a promoter or inhibitor whether it is valid for that role. For an operation $op'$ identified as a promoter, we check whether its post-conditions overlap with the pre-conditions of any of the operations of the root service $s$. If not, $op'$ is removed as a promoting operation. Similarly for an operation identified as an inhibitor, we check whether its post-conditions overlap with the pre-conditions of all of the operations of $s$. If yes, it means that the invocation of $op'$ does not inactivate at least one of $s$'s operations. As a result, we remove it as an inhibiting operation.

### 4.2.2.1 Complexity Analysis

We compare the computation complexity of the static verification algorithm against a naive exhaustive search algorithm. In the exhaustive search algorithm, the time $T_1$ needed to verify whether all the indirect recognition based compositions are statically valid is:

$$T_1 = O[N_{ws} \times N_{oi} \times N_{pout} \times N_{ws} \times N_{oi} \times N_{pin}]$$
$$= O[N_{ws}^2 \times N_{oi}^2 \times N_{pout} \times N_{pin}]$$

The time $T_2$ needed to verify whether all the promotion and inhibition based compositions are statically valid is:

$$T_2 = O[N_{ws} \times N_{oi} \times N_{pout} \times N_{ws} \times N_{oi} + N_{ws} \times N_{oi} \times N_{pin} \times N_{ws} \times N_{oi}]$$
$$= O[N_{ws}^2 \times N_{oi}^2 \times (N_{pout} + N_{pin})]$$

Thus, the total time $T$ needed to verify whether all compositions are valid in the exhaustive approach is:

$$T = O[N_{ws}^2 \times N_{oi}^2 \times (N_{pout} \times N_{pin} + N_{pout} + N_{pint})]$$

We now analyze the complexity of Algorithm 10. Since we rely on ontological index nodes as the recognition medium, the chance $p_1$ that each ontology index node will have an indirect recognition based composition is:

$$p_1 = \frac{N_{ws} \times N_{oi} \times N_{pout}}{|Ont|} \times \frac{N_{ws} \times N_{oi} \times N_{pin}}{|Ont|}$$

The chance $p_2$ that each ontology index node will have a promotion and inhibition based composition is:

$$p_1 = \frac{N_{ws} \times N_{oi} \times (N_{pout} + N_{pin})}{|Ont|} \times \frac{N_{ws}}{|Ont|}$$

Thus, the total time $T$ needed to verify whether all compositions are valid in our approach is:

$$
\begin{aligned}
T &= O[p_1 \times |Ont| + p_2 \times N_{oi} \times |Ont|] \\
&= O[|Ont| \times (\frac{N_{ws}}{|Ont|})^2 \times N_{oi} \times (N_{pout} \times N_{pin} + N_{pout} + N_{pint})] \\
&= O[(\frac{N_{ws}}{|Ont|}) \times N_{ws} \times N_{oi} \times (N_{pout} \times N_{pin} + N_{pout} + N_{pint})]
\end{aligned}
$$

As compared to the exhaustive approach, there is an $|Ont|$ fold improvement in our approach. When the ratio of $\frac{N_{ws}}{|Ont|}$ stays constant, our approach has the performance that is linear to the number of Web services.

#### 4.2.2.2 Complexity Simulation

We conducted experiment to simulate the performance of the static verification algorithm. We used the same runtime environment and configuration variables as shown in Table 4.3. To simulate the pre- and post-conditions, we create a precondition for each input parameter stating that it should fall in a range uniformly selected from [0.0, 1.0). Likewise, we create a postcondition for each output parameter stating that it should fall in a range uniformly selected from [0.0, 1.0). During simulation, we treat an overlap between a postcondition and a precondition as a match. Similar to the simulation for the filtering algorithms, for each pair of $(n_s, s\_num)$, we run through the static verification algorithm 10 times. We then take the average of the total processing time from these runs and plot it in Figure 4.7. From simulation results in both Figures 4.7, we see that for a given services to ontology index nodes ratio, the total verification time is linear to the number of services used as input.

### 4.2.3 Linking

In this section, we describe algorithms for linking composition leads into composition networks. Since a composition lead can start with either an operation (in the case of indirect recognition) or a service (in the case of promotion or inhibition), we split the establishment of link references among these segments into two steps. We show the first step in Algorithm 11.

Within this step, we sweep all composition leads that start with an operation, This step checks, for each lead starting with an operation, all parameter providers for each of the input parameters of the root operation to see if there is a lead that starts with the same operation. If so, it replaces the operation sub node in the current

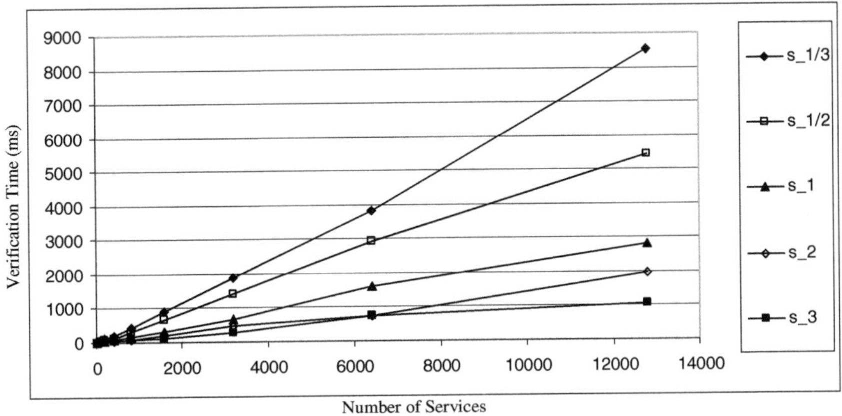

Fig. 4.7 Verification Complexity Simulation: Processing Time vs. Number of Services

---

**Algorithm 11** Composition Linking Sweep1

---

**Input**: Leads of compositions $L_{cs}$
**Output**: Leads of composition $L_{cs}$ marked with link references among operations.
**Variables**: composition leads $l_{cs}$ and $l'_{cs}$, parameter subNode *para*, and operation subNode *op*.

```
 1: for all l_cs ∈ L_cs do
 2:    if l_cs.rootNode is an operation then
 3:       for all parameter subNode para of l_cs.rootNode do
 4:          for all provider operation subNode op of para do
 5:             if ∃l'_cs ∈ L_cs : op = l'_cs.rootNode then
 6:                replace op with a reference to l'_cs under para;
 7:                mark l'_cs as being referenced;
 8:             end if
 9:          end for
10:       end for
11:    end if
12: end for
```

---

tree with the root node of the newly found lead. In addition, the algorithm marks the newly found lead as being referenced so that it won't get included again later as a different lead since it is already part of the current lead.

In the second step shown in Algorithm 12, we first sweep all composition leads that start with a service (lines 1-16) and then all composition leads that start with an operation (lines 17-25). During the sweep for each of the leads that starts with a service, this step first traverses to the sub node representing the entity providing the service and then checks, for each of the entity node's sub nodes representing an operation that either promotes the service (via outputting the same entity as a parameter) or inhibits the service (via taking as an input parameter the same entity), whether there is a lead that starts with the same operation. If so, it replaces the op-

---

**Algorithm 12** Composition Linking Sweep2

---

**Input**: Leads of composition segments $L_{cs}$ marked with link references among operations.
**Output**: Lead composition networks $L_c$.
**Variables**: Composition leads $l_{cs}$ and $l'_{cs}$, entity $e$, operation subNode $op$, and provider service subNode $s$ of $op$.

1: **for all** $l_{cs} \in L_{cs}$ **do**
2:     **if** $l_{cs}.rootNode$ is a service **then**
3:         $e \leftarrow$ entity subNode of $l_{cs}.rootNode$
4:         **for all** operation subNode $op$ of $e$ **do**
5:             **if** $\exists l'_{cs} \in L_{cs} : op = l'_{cs}.rootNode$ **then**
6:                 replace $op$ with a reference to $l'_{cs}$ under $e$;
7:                 mark $l'_{cs}$ as being referenced;
8:             **end if**
9:             $s \leftarrow$ service subNode of $op$
10:            **if** $\exists l'_{cs} \in L_{cs} : s = l'_{cs}.rootNode$ **then**
11:                replace $s$ with a reference to $l'_{cs}$ under $op$
12:                mark $l'_{cs}$ as being referenced;
13:            **end if**
14:         **end for**
15:     **end if**
16: **end for**
17: **for all** $l_{cs} \in L_{cs}$ **do**
18:     **if** $l_{cs}.rootNode$ is an operation **then**
19:         $s \leftarrow$ service subNode of $op$
20:         **if** $\exists l'_{cs} \in L_{cs} : s = l'_{cs}.rootNode$ **then**
21:             replace $s$ with a reference to $l'_{cs}$ under $op$;
22:             mark $l'_{cs}$ as being referenced;
23:         **end if**
24:     **end if**
25: **end for**
26: **for all** $l_{cs} \in L_{cs}$ **do**
27:     **if** $l_{cs}$ is not referenced **then**
28:         Add $l_{cs}$ to $L_c$;
29:     **end if**
30: **end for**

---

eration sub node with the root node of the newly found lead, which is then marked as being referenced. The algorithm then traverses to the operation's sub node representing the service that provides this operation and checks to see if there exists a lead that starts with the same service. If so, it replaces the service sub node with the root node of the newly found lead, which is then marked as being referenced. Next, we need to ensure that the service sub node of each of the leads that starts with an operation is also properly updated with the root node of a lead that starts with the same service, if such a lead exists. This is done by sweeping each lead that starts with an operation, traversing to its sub node representing the service that provides the operation, checking to see if there exists a lead that starts with the same service, and if so replacing the service sub node with the root node of the newly found lead and marking the lead as being referenced. Finally, each of the composition leads, which may have been extended via references to other composition leads, is added to the collection of lead composition networks if it is not marked as being referenced.

### 4.2.3.1 Complexity Analysis

We now analyze the complexity of our linking algorithms. In Algorithm 11, line 1 is an operation that is linear to $N_{ws}$. Line 3 has a factor of $\frac{N_{ws}}{|Ont|} \times N_{pin}$. Line 4 has a factor of $\frac{N_{ws}}{|Ont|} \times N_{pout}$. Line 5 has a factor of $logN_{ws}$. In Algorithm 12, lines 1, 17 and 26 all have a factor of $N_{ws}$. Line 4 has a factor of $\frac{N_{ws}}{|Ont|} \times (N_{pin} + N_{pout})$. Lines 5, 10 and 20 all have a factor of $logN_{ws}$. Thus, the total time needed for both Algorithms 11 and 12 is:

$$
\begin{aligned}
T &= O[N_{ws}\{\frac{N_{ws}}{|Ont|} \times N_{pin} \times \frac{N_{ws}}{|Ont|} \times N_{pout} \times logN_{ws} \\
&\quad + \frac{N_{ws}}{|Ont|} \times (N_{pin} + N_{pout}) \times logN_{ws} + logN_{ws} + 1\}] \\
&= O[N_{ws}\{(\frac{N_{ws}}{|Ont|})^2 \times N_{pin} \times N_{pout} \times logN_{ws} \\
&\quad + \frac{N_{ws}}{|Ont|} \times (N_{pin} + N_{pout}) \times logN_{ws} + logN_{ws} + 1\}]
\end{aligned}
$$

If we keep $\frac{N_{ws}}{|Ont|}$ constant, then $T = O[N_{ws}logN_{ws}]$.

### 4.2.3.2 Complexity Simulation

We conducted experiment to simulate the performance of the linking algorithms. We used the same runtime environment and configuration variables as shown in Table 4.3. Similar to the simulation for the filtering and static verification algorithms, for each pair of $(n_s, s\_num)$, we run through the static verification algorithm 10 times. We then take the average of the total processing time from these runs and plot it in Figure 4.8. Although the performance is not linear to $N_{ws}$, as is that for filtering and static verification, the overall time it takes to complete the linking phase is relatively much shorter than are those for the other two types of algorithms given the range of number of services that is of interest here.

## 4.2.4 Effects of Operation Diversity and Complementarity

Among the Web services composition leads, many may turn out to be useless or frivolous. In drug discovery, frivolous leads can consume most of the screening

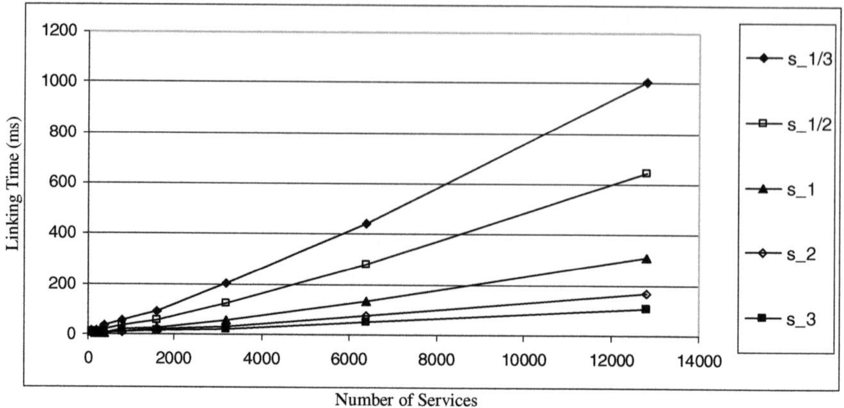


**Fig. 4.8** Linking Complexity Simulation: Processing Time vs. Number of Services

time. An effective way to tackle this problem is to increase the chemical diversity of the compound library [103]. Likewise, the diversity of web service operations plays an important role in Web service mining. We define operation diversity below. *Operation Diversity* of mining context $C$ is defined as:

$$diversity(C) = \frac{Num(node_{ont})}{\sum_{Op \in C} Num(op.parameters)} \qquad (4.7)$$

where

- $Num(node_{ont})$ is the number of nodes in the ontology that are referred to by operation parameters of operations in $C$, and
- $Num(op.parameters)$ is the number of $op$'s parameters including both its input and output parameters. $\Diamond$

Web service operation diversity has a direct effect on Web service screening. If operations are not diverse enough, we may have too many operations crowding over just a few ontology nodes, and consequently end up with too many frivolous composition leads that can overburden the evaluation phase. If operations are too diverse, we may have too few operations using any given ontology nodes, and have no composition lead after all. However, the increased difficulty of producing operation recognitions due to a more diverse operation distribution also increases the value of such recognition. While diversity indicates the value of recognition, the following measure gives a good indication of the probability of recognition occurrences. *Operation Complementarity* is defined as:

$$complementarity(C) = \frac{Num_{ps}(node_{ont})}{Num(node_{ont})} \qquad (4.8)$$

where

- $Num_{ps}(node_{ont})$ is the number of nodes in the ontology that have both publishers and subscribers from $C$, and
- $Num(node_{ont})$ is the number of nodes in the ontology that are referred by parameters of operations in $C$. $\diamond$

If we use a group of relevant leads (GRL) to refer to leads that center around the same completely bound operation, then complementarity serves as an indicator of either the population or the density of GRLs that will result from Web service screening. Obviously as complementarity approaches 1, more operations within the mining context become completely bound. If this trend is also coupled with a high value of diversity, then more GRLs are expected to result from the mining process.

## 4.3 Post-screening Analysis and Evaluation

Not all service compositions discovered during the screening phase are necessarily interesting and useful. The purpose of post-screening analysis and evaluation is to identify those that are truly interesting and useful. In this chapter, we cover four distinct steps in achieving this: *objective evaluation, interactive hypothesis formulation, simulation* and *subjective evaluation*.

### 4.3.1 Objective Evaluation

Objective evaluation aims at using objective measures to evaluate the interestingness and usefulness of composed Web services. In this section, we first introduce the concepts of operation similarity, domain correlation and unrelatedness. These are used in our objective measures for interestingness and usefulness, which are described thereafter. We then discuss objective evaluation strategies for the purpose of narrowing down the candidate pool for further consideration.

#### 4.3.1.1 Operation Similarity

The concept of operation *similarity* is relevant when we study the interestingness of an *indirect recognition*-based composition. The similarity of two operations can be measured by comparing their input parameter set, output parameter set, pre-conditions and post-conditions. We use the following function to measure the similarity between $op_i$ and $op_j$:

$$Sim(op_i, op_j) = c_p \left( \frac{|P_{in}(op_i) \cap P_{in}(op_j)|}{|P_{in}(op_i) \cup P_{in}(op_j)|} \times \frac{|P_{out}(op_i) \cap P_{out}(op_j)|}{|P_{out}(op_i) \cup P_{out}(op_j)|} \right)$$
$$+ c_c \left( \frac{|C_{pre}(op_i) \cap C_{pre}(op_j)|}{|C_{pre}(op_i) \cup C_{pre}(op_j)|} \times \frac{|C_{post}(op_i) \cap C_{post}(op_j)|}{|C_{post}(op_i) \cup C_{post}(op_j)|} \right) \quad (4.9)$$

where $c_p$ and $c_c$ are weights such that $0 \le c_p, c_c \le 1$ and $c_p + c_c = 1$. $|P|$ and $|C|$ give the size of parameter set $P$ and condition set $C$, respectively. According to Eq. (4.9), $Sim(op_i, op_j)$ ranges from 0 to 1, with 1 indicating that the two operations have the same parameters and conditions.


### 4.3.1.2 Domain Correlation and Unrelatedness

Domain correlation $\Gamma$ measures the relevancy of two domains $d_i$ and $d_j$ or the cohesion of the same domain (when $i = j$). It is defined as:

$$\Gamma[d_i, d_j] = e^{-\frac{1}{\varepsilon_0(n+1)}} \quad (4.10)$$

where $n$ is the number of unique pairs of operations, $\{(op_i, op_j) \mid op_i \in d_i, op_j \in d_j\}$, that are previously known to have been involved in a composition. When $n = 0$, the correlation between two domains in Eq. (4.10) is assigned an initial value of $\gamma_0 = e^{-\frac{1}{\varepsilon_0}}$. Eq. (4.10) shows that $\Gamma$ for two domains quickly approaches 1 as $n$ increases.

We define the multiplicative inverse of the domain correlation as *domain unrelatedness*, $\Theta$. We bound the maximum value of $\Theta$ to 1, thus

$$\Theta[d_i, d_j] = \frac{\gamma_0}{\Gamma[d_i, d_j]} \quad (4.11)$$


### 4.3.1.3 Objective Interestingness

*Interestingness* indicates how interesting a Web service composition discovered from the screening phase really is. We consider two types of application area where this needs to be determined. The first is e-government and e-commerce, where QoWS plays an important role in the case of *direct recognition*-based composition. We say the composition is interesting if it exhibits better qualities than all previously discovered similar operations. The second application area is biological pathway discovery, where the significance of QoWS is less important as we don't expect any occurrences of *direct recognition*-based composition. In this section, we attempt to devise an objective interestingness measure factoring in various considerations for the other three recognition mechanisms that are applicable to both application areas. We define interestingness, $I$, as a tuple $[A, N, U_q, D_v, S, W]$, where

- $A$ is the *actionability* of the composition
- $N$ is the *novelty* of the composition,

- $U_q$ is the *uniqueness* of the composition,
- $D_v$ is the *diversity* of the composition constituents,
- $S$ is the *surprisingness* of the composition,
- $W$ is the product of expert-assigned weights $w_i$ ($w_i > 1$) to domain ontology index nodes that are involved in a composition. We choose to multiply all such weights involved in a composition to reflect their subjective interestingness-enhancing effect.

*Actionability*. We define actionability as a binary (i.e., 1 for actionable, 0 for non-actionable) representing whether the composability of a composition can be verified through simulation or live execution. A non-actionable composition is considered uninteresting. Thus actionability contributes multiplicatively towards the overall interestingness.

*Novelty*. We define novelty as a binary (i.e., 1 for novel, 0 for old or known). The source of this information may be a database or registry that keeps track of known service compositions or, in the case of pathway discovery, a pathway base. An old or known composition is considered uninteresting. Thus novelty also contributes multiplicatively towards the overall interestingness.

*Uniqueness*. Uniqueness, $U_q$, measures how unique a composition is. We use the following function to calculate uniqueness:

$$U_q = \begin{cases} 1, & \textit{promotion or inhibition} \\ 1 - Max_{op \in D} Sim(comp(OP_s, op_t), op), & \textit{indirect recognition} \end{cases} \quad (4.12)$$

For both *promotion* and *inhibition*, the uniqueness is set to 1 due to the validity of the composition. For *indirect recognition*, $D$ is a reference set of domains. Obviously the more similar the composed operation is to an existing operation, the less unique it is regarded. $U_q$ can thus vary between 0 and 1.

*Diversity*. Diversity $D_v$ indicates how diverse components involved in a Web service composition are. We use the following objective function to measure diversity:

$$D_v = \begin{cases} \Sigma_{d_s \in D(s)} \Theta[d(op), d_s], & \textit{promotion or inhibition} \\ \Sigma_{op \in OP_s} \Theta[d(op), d(op_t)], & \textit{indirect recognition} \end{cases} \quad (4.13)$$

If we replace $\Theta$ with $\Gamma$, Eq. (4.13) can be rewritten as:

$$D_v = \begin{cases} \Sigma_{d_s \in D(s)} \frac{\gamma_0}{\Gamma[d(op), d_s]}, & \textit{promotion or inhibition} \\ \Sigma_{op \in OP_s} \frac{\gamma_0}{\Gamma[d(op), d(op_t)]}, & \textit{indirect recognition} \end{cases} \quad (4.14)$$

For *promotion* or *inhibition* involving $op$ and $s$, $D(s)$ is a set of domains $s$ is involved with. In all cases, $d()$ provides the domain of a given service operation. According to Eq. (4.14), in the case of promotion or inhibition, the more domains $s$ is involved with and the less the correlation between these domains and that for $op$, the higher the value of diversity. In the case of indirect recognition, both the number and the domains of source operations $op \in OP_s$ contribute to the overall diversity. The more source operations that are involved in the composition and the less the correlation between their domains and the domain of the target operation, the higher the value of diversity.

***Surprisingness***. Surprisingness $S$ indicates how unexpectedly a Web service composition is achieved. We use the following objective function to measure it:

$$S = \begin{cases} max_{d_s \in D(s)} \Theta[d(op), d_s], & \textit{promotion or inhibition} \\ max_{op \in OP_s} \Theta[d(op), d(op_t)], & \textit{indirect recognition} \end{cases} \tag{4.15}$$

Replacing $\Theta$ with $\Gamma$, Eq. (4.15) becomes:

$$S = \begin{cases} \dfrac{\gamma_0}{min_{d_s \in D(s)} \Gamma[d(op), d_s]}, & \textit{promotion or inhibition} \\ \dfrac{\gamma_0}{min_{op \in OP_s} \Gamma[d(op), d(op_t)]}, & \textit{indirect recognition} \end{cases} \tag{4.16}$$

Unlike diversity, the measure of surprisingness focuses on the correlation between two domains having the biggest contribution to the overall value. According to Eq. (4.16), a composition achieved by combining two services from domains previously known to be very relevant is less surprising than one from those previously known to be less relevant. Due to the similarity between Eq. (4.14) and Eq. (4.16), it turns out that surprisingness and diversity are correlated. As our results in Section 4.3.1.6 shows, such correlation also varies as the number of service operations increases.

### 4.3.1.4 Objective Usefulness

*Usefulness* is the next determination criteria following *interestingness*. We mentioned in Section 3.4 that the meaning of usefulness is sometimes relative and subjective. However, when the field of study and domain of interest are narrow enough, we often have a good general consensus on what to expect of usefulness. For example, in the field of drug discovery, usefulness of drugs is manifested by their capability to inhibit or remedy abnormal receptors as they bind with drug molecules. In other words, the composition of the abnormal receptors and drug molecules is useful because it results in some new properties that are responsible for the improved overall interaction of the receptors with the surrounding cells. As a result, the spread of the disease inside the body is either controlled or stopped. In another example involving the conference Web service (CWS) introduced in Section 3.2, instead of requiring conference participants to find and reserve hotel rooms and ground transportation by searching various travel services individually, the composite service CWS can provide a one-stop service point that not only guarantees the synchronization of the participants' schedule with that of the conference, but also offers them a discount rate through its partnership with the selected travel service.

While it is difficult to objectively quantify the ramification of new properties emerged out of a service composition using a simple usefulness measure, we take into consideration elements of usefulness that can be objectively evaluated for the cases of direct and indirect recognition. Due to the prevalent use of QoWS attributes in e-commerce and e-government, usefulness in these settings could be calculated for the following two cases:

1. There are multiple operations competing to provide each of the input parameters of a completely bound operation within a group of relevant leads (GRL) centered around an operation interface. Each of these competing operations may in turn have implementations provided by multiple services. If the overall composition provides the same or similar function not seen before, then the usefulness can be at least partially calculated using the overall QoWS. The purpose of doing that is so that all the candidate compositions can then be compared to find out which one of them exhibits the highest QoWS.
2. If a composition provides a function that is the same as or similar to an existing function, usefulness could be expressed through its QoWS improvement achieved over the existing function.

In either case, QoWS-based usefulness can be determined either statically if service QoWS are registered over time and become wildly known or measured at runtime if they are vague and have to be determined dynamically.

#### 4.3.1.5  Evaluation of Objective Interestingness and Usefulness

When multiple compositions are identified through the screening phase, the candidate pool for further consideration can be reduced through selecting service compositions that exhibit the highest values of objective interestingness and/or usefulness. This can be done using two approaches, namely *weighted function* and *skyline* methods. We describe the two approaches in this section.

**Objective Function**

An object function taking into account interestingness measures can be devised as follows:

$$I = AN(c_u U_q + c_d D_v + c_s S) \prod_{i=1}^{m} w_i \qquad (4.17)$$

where $c_u$, $c_d$ and $c_s$ are weights for the weighted function such that $0 \leq c_u, c_d, c_s \leq 1$ and $c_u + c_d + c_s = 1$. $m$ is the number of expert-assigned weights $w_i$ ($w_i > 1$) to operation interfaces and domain ontology index nodes that are involved in a composition. We choose to multiply all such weights involved in a composition to reflect their subjective interestingness-enhancing effect. Likewise, a similar weighted function on usefulness in the e-commerce environment can be calculated using:

$$
U = \begin{cases}
\sum_{q \in pos} w_q \frac{q - q_{min}}{q_{max} - q_{min}} + \sum_{q \in neg} w_q \frac{q_{max} - q}{q_{max} - q_{min}}, & \text{\textit{new function (case 1)}} \\[2ex]
\sum_{q \in pos} w_q \frac{\delta q}{q_{max} - q_{min}} - \sum_{q \in neg} w_q \frac{\delta q}{q_{max} - q_{min}}, & \text{\textit{existing function (case 2)}}
\end{cases}
$$

$$(4.18)$$

where *pos* is a set of quality attributes (e.g., reliability) that contribute positively towards the weighted function while *neg* is a set of quality attributes (e.g., response time) that contribute negatively towards the weighted function. $w_q$ are weights assigned by users to each quality attributes such that $w_q \geq 0$ and $\sum w_q = 1$. $q$ represents the value of an aggregate lead QoWS attribute calculated according to Table 3.1. $\delta q = q_{composition} - q_{existing}$. $q_{max}$ and $q_{min}$ are respectively the maximum and minimum values of the same attribute among all leads in the GRL. The first half of Eq. (4.18) is essentially a combined form of the *scaling phase* and *weighting phase* proposed in [107]. The second half of Eq. (4.18) extends the function to the improvement of QoWS calculated between two compositions.

Such an weighted function, once configured with all the weights, can be rather simple to use. Unfortunately, the configuration itself requires the user to first express his/her preferences over several interestingness-related measures or quality attributes as numeric weights. The specification of these weights could be rather demanding yet arbitrary due to the uncertain nature of transforming user preferences to numeric weights. Often such preferences could be rather vague before the data are being presented to the user. As a result, the user may have to go through a time consuming trial-and-error process, as the data are being presented, to arrive at a desired combination of such weights. In addition, no matter how the weights are distributed along different measures, there is always a possibility that some of the compositions having high values in certain measures may be suppressed due to their low values in others. This is especially true when the high values are associated with small weights and the low values are associated with big weights. This could be a rather undesirable situation if the user wants to include compositions with a very high value in any measure.

## Skyline

Another popular approach in selecting service compositions that exhibit high values of desired properties is the use of *skyline* operator, which originates from the database community [29]. A skyline is defined as a set of objects that are not *dominated* by other objects. In a multi-objective environment such as those listed in Eq. (4.17), composition $comp_a$ dominates composition $comp_b$ if $comp_a$ exhibits better value in at least one dimension than does $comp_b$ and values as good as or better than does $comp_b$ in all other dimensions. The skyline operator addresses well the problem faced by the weighted objective function as the user is not required to come up with an optimal combination of all the weights used in the weighted function.

### 4.3.1.6 Interestingness Simulation

In our simulation, we focus on investigating the interestingness skyline of service compositions as skylines of QoWS based measures have already been extensively studied in Yu's work [106]. In addition, we focus on the study of interestingness of compositions obtained through indirect recognition since they require more computation according to Eqs. (4.12), (4.14) and (4.16). Table 4.4 lists the configuration variables used in our experiment.

**Table 4.4** Experiment Settings for Interestingness Simulation

| Variable | Value or Range |
|---|---|
| Domain ontology index nodes | 50000 |
| Ratio of ontology index nodes assigned user weight | 0.002 |
| Range of user weight | $1.0 - 5.0$ |
| Number of domains | 50 |
| Number of operation interfaces per domain $n_o$ | $50 - 500$ |
| Input parameters per operation $n_p$ | $0 - 5, 0 - 10$ for Figure 4.10 (c) |
| Output parameters per operation | $0 - 5$ |
| Pre/Post-condition range (float) | $0 - 1.0$ |
| Similarity $c_c, c_p$ | $0.4, 0.6$ |
| $\varepsilon_0$ | $0.4, 0.5, 0.6$ |

Since both actionability and novelty are boolean variables, we ignore these two variables in our simulation. During each iteration of our mining algorithms, we pick a value for the number of operation interfaces per domain and use that to populate operations in 50 domains. For each domain operation, we generate its input/output parameters such that the number of these parameters uniformly falls in the range of 0 to 5. Each of these parameters is associated with a Domain Ontology Index Node (DOIN), which is identified with a sequence number. For simplicity, we flatten all the DOINs (i.e., no *inheritance* and *composition* relationships among ontology nodes) so that only exact matches and synonyms will be considered. We place these DOINs (50000 of them for the experiment) in a circular buffer so that the last sequence number is next to the first one. To study the contribution of user assigned weights on DOINs towards the interestingness, we randomly choose 100 nodes ($50000 \times 0.002$) using a uniform distribution and assign a weight uniformly distributed between 1.0 and 5.0. To simulate the cohesive nature of DOINs in a domain, we pick them for the domain using a Gaussian distribution around a mean sequence number randomly chosen for the domain according to a uniform distribution. We assume that each parameter has an equal chance of being associated with a DOIN. To simulate the pre- and post-conditions, each parameter is symbolically given a range randomly chosen between 0 and 1.0 using a uniform distribution. We use the overlap of two such ranges (see Eq. (4.9)) to calculate the contribution of these conditions towards the similarity of two operations. During each mining iteration based on the chosen number of operation interfaces per domain, $n_o$, we calculate uniqueness, diversity, surprisingness and weight product for compositions discovered in the iteration. These values are then normalized using the following equation:

$$v = \frac{v - v_{min}}{v_{max} - v_{min}} \tag{4.19}$$

The interestingness skylines can be shown as a surface formed in a 4D space with compositions' uniqueness $U_q$, diversity $D_v$, surprisingness $S$ and weight product $\prod_{i=1}^{m} w_i$ as the coordinates. However, in order to plot them, we decide to represent each as a pair of 3D surfaces using the following pair of coordinates: $\{U_q, D_v, \prod_{i=1}^{m} w_i\}$ and $\{U_q, S, \prod_{i=1}^{m} w_i\}$. Figure 4.9 uses blue circles to highlight skyline points. It shows the pair of 3D interestingness skylines for different numbers of operation interfaces per domain. We see that as the number of operation interfaces per domain increases, the number of discovered compositions also increases dramatically. However, the interestingness skyline keeps a population of top candidates with a relatively stable size.

Our experiment results also reveal a strong correlation between diversity and surprisingness. To take a closer look at this, we plot the pairs of (*diversity*, *surprisingness*) for each experiment setting using a 2-D graph. Figure 4.10 (a) shows two of these plots selected based on varying values of $n_o$. At the top of each plot is information presented in the form of $n_s$-$N_c$, where $n_s$ is the number of source operations involved in a composition and $N_c$ is the number of composition instances containing $n_s$ source operations. $n_s$ ranges from 1 to 5, corresponding to the the number of input parameters an operation can have (see Table 4.4). Figure 4.10 (a) shows that as $n_o$ increases, distribution of compositions shifts towards higher values of $n_s$, indicating an increasing likelihood for a composition to involve more source operations. For example, when $n_o = 50$, there are no composition for $n_s = 4$ or 5. When $n_o = 450$, the number of compositions for $n_s = 5$ reaches 1680, almost half of the number of those for $n_s = 1$. Each of the five distinct distribution lines corresponds to a value of $n_s$.

For illustration purposes, we choose to plot the dimension of surprisingness using a log scale. Consequently, an interesting correlation pattern between diversity and surprisingness emerges. As the normalized value of surprisingness approaches 0, the distribution lines start to become parallel. In addition, these distribution lines are evenly spaced out with a distance of around 0.1. Based on Eq. (4.14), since the value chosen for surprisingness is the largest element used for calculating diversity, as the normalized value of surprisingness approaches 0, all such elements approaches $e^{-\frac{1}{\varepsilon_0}}$, thus the distribution lines are evenly spaced out. The analytical value of the spacing distance can be calculated using the following:

$$distance = \frac{e^{-\frac{1}{\varepsilon_0}}}{x + (n-1)e^{-\frac{1}{\varepsilon_0}}} \tag{4.20}$$

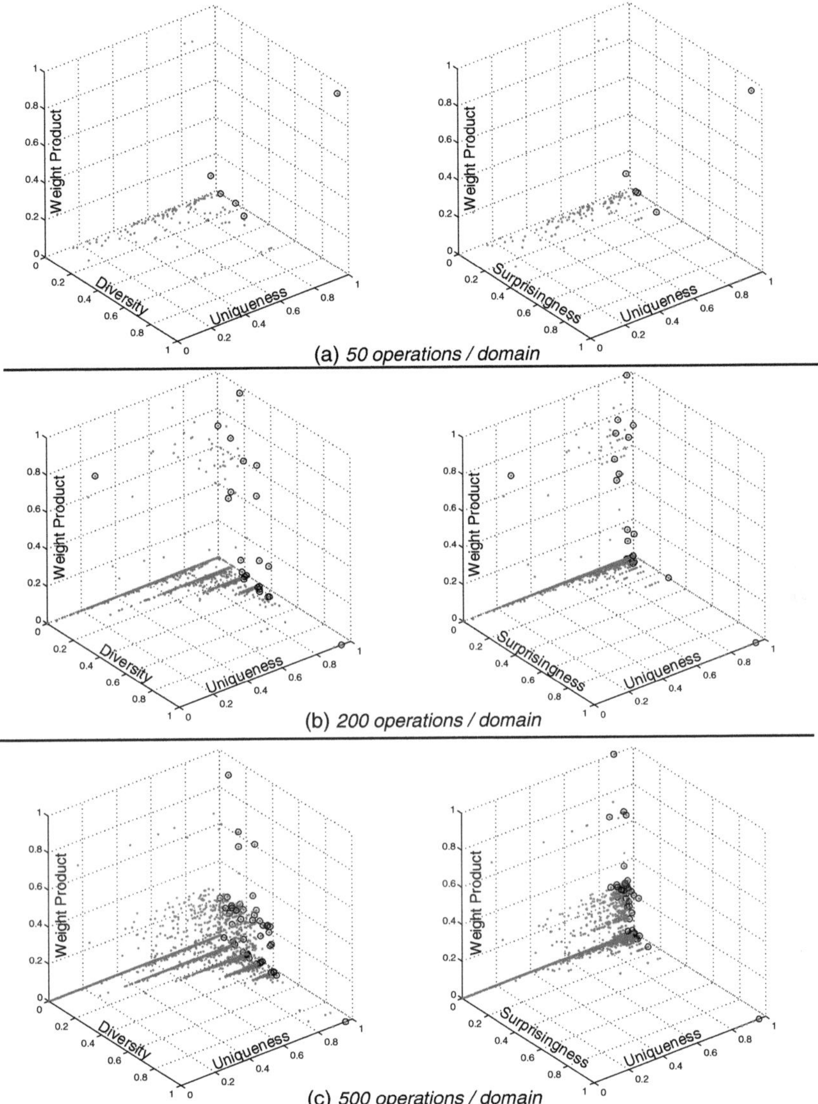



**Fig. 4.9**  Skylines vs. Number of Operations

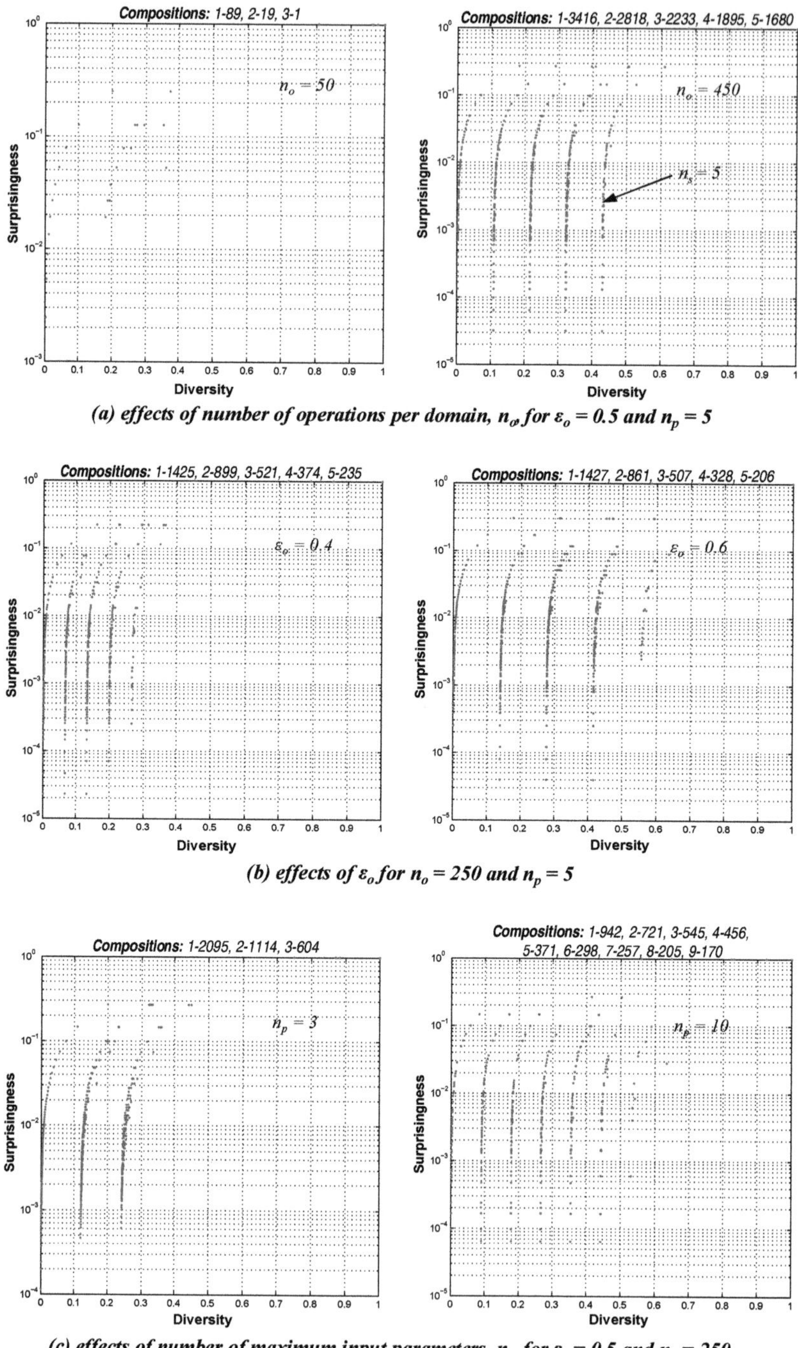


*(a) effects of number of operations per domain, $n_o$, for $\varepsilon_o = 0.5$ and $n_p = 5$*

*(b) effects of $\varepsilon_o$ for $n_o = 250$ and $n_p = 5$*

*(c) effects of number of maximum input parameters, $n_p$, for $\varepsilon_o = 0.5$ and $n_o = 250$*

**Fig. 4.10** Correlation between Diversity and Surprisingness

where $x$ is the largest element of diversity and should fall in the range of $(e^{-\frac{1}{\varepsilon_0}}, 1]$. $n$ is the maximum number of input parameters for an operation. Thus the range of the above distance for $\varepsilon_0 = 0.5$ and $n = 5$ should be:

$$range = (\frac{e^{-\frac{1}{\varepsilon_0}}}{1+4e^{-\frac{1}{\varepsilon_0}}}, \frac{e^{-\frac{1}{\varepsilon_0}}}{5e^{-\frac{1}{\varepsilon_0}}}] = (0.088, 0.2] \qquad (4.21)$$

Figure 4.10 (b) shows two plots with varying values of $\varepsilon_0$. As $\varepsilon_0$ increases, the spacing distance between distribution lines also increases. Using Eq. (4.21), we see that for $\varepsilon_0 = 0.4$, the lower bound of Eq. (4.21) becomes 0.062. This lower bound increases to 0.108 for $\varepsilon_0 = 0.6$. Obviously, when $\varepsilon_0$ is small, domains are more cohesive and less correlated, resulting in smaller values of the overall diversity. As $\varepsilon_0$ increases, domains become less cohesive and more correlated, resulting in higher values of the overall diversity.

Figure 4.10 (c) shows two plots with varying values of the number of maximum input parameters, $n_p$. Given $\varepsilon_0 = 0.5$, the range of spacing distance shifts from $(0.107, 0.333]$ for $n_p = 3$ to $(0.075, 0.143]$ for $n_p = 7$, and to $(0.061, 0.1]$ for $n_p = 10$, respectively. Note that as the number of maximum input parameters increases, more distribution lines emerge. However, the spacing distribution among these lines reduces. Both Figures 4.10 (b) and (c) indicate that diversity provides a finer distinction mechanism than surprisingness and can be used to better separate out the most interesting compositions.

## *4.3.2 Interactive Hypothesis Formulation*

To help user formulate hypotheses that would lead to the identification of ultimately useful service compositions or pathways, we have developed strategies aiming at providing user with visual aids towards that goal. We describe these strategies in this section and focus our discussion on relevant strategies used in the discovery of useful pathways.

### 4.3.2.1  Identification of Interesting Segments in a Composition Network

After a service composition or pathway network is discovered from the screening phase, the identification of interesting segments within the network would help user focus more towards them as they are expected to become part of the final outcome from the service mining process. The evaluation strategies discussed in Section 4.3.1.5 can be used in general to identify service compositions of high interestingness. In practice, to enable the automatic identification of interesting segments within a pathway network, we also had to extend each WSML service modeling a biological process to declare the source information in its nfp section. This information allows the identification of novel edges between service models in a discovered pathway network. The identification relies on comparing the source indicator of edges in the pathway graph representing three types of service/operation recognitions (Section 3.3): *promotion*, *inhibition*, and *indirect recognition*. These edges can then be automatically highlighted in the graph.


### 4.3.2.2 Establishment of Connected Graph

Once interesting edges in a service composition or pathway network are highlighted to the user, he/she can then use them as hints in selecting nodes of interest for further exploration. We have developed the following strategy [111] to link both interesting edges and user selected nodes into a fully connected graph to the extent possible:

1. Coalescing nodes (e.g., *a*, *b*, *c* in Figure 4.11) linked by interesting edges into a group.
2. Converting interesting nodes (e.g., *t* picked by user) and groups encompassing interesting nodes (e.g., *c*, *f*) into nuclei, i.e., graph expansion focus nodes.
3. Incrementally expanding all the nuclei. We use the heuristics of connecting all the interesting nodes using as many interesting edges as possible. To achieve this, whenever a newly encountered node is part of a non-nucleus group (e.g., one that contains *h*, *i* and *j*), an additional expansion is also triggered and the whole group are engulfed. The expansion stops when all nuclei are connected or when all nodes in the graph are visited.

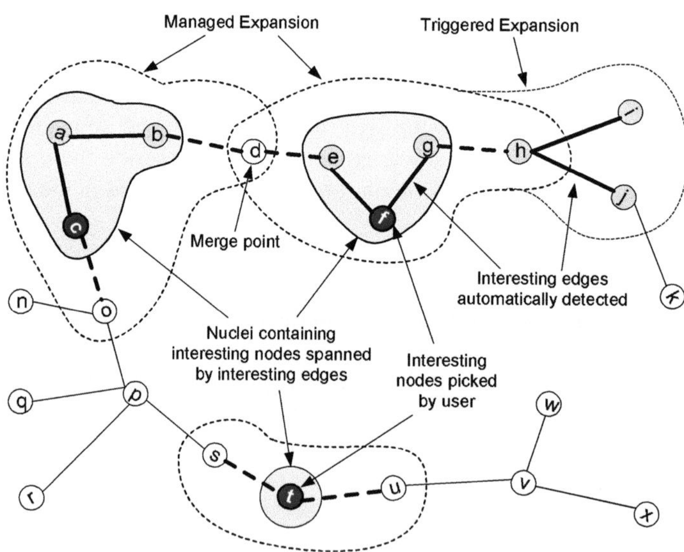


**Fig. 4.11** Expansion of Interesting Segments in Composition Graph

---

**Algorithm 13** Coalesce Interesting Nodes

---

**Input**: Global edge set $E_a$ in composition graph, global node set $N_a$ in composition graph
**Variables**: Global edge references set $E_b$, global node reference set $N_b$, number of interesting node references $N_i$, interesting edge reference set $S_e$, group node reference set $S_n$, and group edge reference set $T_e$

1: create $E_b$;
2: create $N_b$;
3: calculate $N_i$ from $N_b$;
4: **if** $N_i \leq 1$ **then**
5:     return;
6: **end if**
7: move all interesting edges in $E_b$ into $S_e$;
8: **while** $\exists e \in S_e$ **do**
9:     create $S_n$;
10:     create $T_e$;
11:     **coalesce**$(e, S_e, T_e, S_n)$;
12:     remove first node $n_g$ found in $S_n$;
13:     convert $n_g$ to a group node by attaching $S_n$ to $n_g$;
14:     **if** $S_n$ is marked interesting **then**
15:         **for all** $e \in T_e$ **do**
16:             mark as connected the edge in $E_a$ that $e$ refers to;
17:         **end for**
18:         convert $n_g$ to a nucleus node by setting $n_g.nucleus = n_g$ and $n_g.distance = 0$;
19:     **end if**
20:     **while** $\exists n \in S_n$ **do**
21:         **for all** $e$ connected to $n$ **do**
22:             associate $e$ with $n_g$ if it does not connect $n_g$;
23:             **if** $e \in E_b$ and thus is not interesting **then**
24:                 **if** one end of $e$ connects to $n$ such that $n \in S_n$ **then**
25:                     redirect that end of $e$ to $n_g$;
26:                 **end if**
27:             **end if**
28:         **end for**
29:         remove $n$ from $S_n$;
30:         remove $n$ from $N_b$;
31:     **end while**
32: **end while**
33: convert all interesting nodes that are not group nodes into nucleus nodes;

---

Connected graphs identified using the above process are then presented to the user as basis for hypothesis formulation. We list algorithm used to achieve steps 1 and 2 in Algorithm 13. We first construct two global reference sets: $E_b$ for edges and $N_b$ for nodes (lines 1 and 2). Since we are trying to connect all interesting nodes, the algorithm stops when $N_i \leq 1$ (lines 4 to 6). The rest of the algorithm aims at coalescing each group of nodes in $N_b$ linked by interesting edges into one node. We first construct $S_e$ (line 7) to initially contain all the interesting edge references. For each remaining edge $e$ picked from $S_e$ (line 8), we construct a group node reference set $S_n$ (line 9) and group edge reference set $T_e$ (line 10). A *coalesce*() function listed in Algorithm 14 is then invoked to coalesce nodes that $e$ connects. In Algorithm 14, we first move $e$ from the interesting edge reference set $S_e$ to the corresponding

---

**Algorithm 14** Coalescing Group Nodes

---

$coalesce(e, S_e, T_e, S_n)$
**Input**: Global edge reference set $N_b$
**Variables**: Number of interesting node reference sets $N_i$

```
 1: move e from S_e to T_e;
 2: for all n connected by e such that n ∈ (N_b − S_n) do
 3:     S_n.add(n);
 4:     if n is interesting then
 5:         if S_n is interesting then
 6:             N_i- -;
 7:         else
 8:             mark S_n as interesting;
 9:         end if
10:     end if
11:     for all e' ∈ S_e such that e' connects n do
12:         coalesce(e', S_e, T_e, S_n);
13:     end for
14: end for
```

---

group edge reference set $T_e$ (line 1). Then for each node $n$, which is in $N_b$ but not $S_n$ and which $e$ connects to (line 2), we add it into the corresponding group node reference set $S_n$ (line 3). If $n$ is an interesting node (line 4) and $S_n$ is already marked as interesting (line 5), then we know that $n$ is not the first interesting node in $S_n$, thus we can reduce the number $N_i$ of interesting node reference sets by 1 (line 6). If $S_n$ is not yet marked as interesting, we need to simply do so (line 8). We then recursively invoke Algorithm 14 for all other edges in $S_e$ that are connected to $n$ (lines 11 to 13). It is conceivable that $S_e$, $T_e$ and $S_n$ may all change as a result of this coalescence process. Going back to Algorithm 13, code in lines 12 through 31 aims at picking a node from each group as the proxy for the whole group during the incremental expansion phase (step 3). To achieve this, we pick out the first node $n_g$ found in $S_n$ (line 12) and convert it to a group node (line 13). Lines 14 through 19 converts $n_g$ to a nucleus node and marks corresponding edges in the global edge set $E_a$ as already connected. Lines 21 through 31 makes $n_g$ a surrogate node for all the other nodes in the same group. In line 33, we also convert interesting nodes (e.g., $t$ in Figure 4.11) that are not group nodes into nucleus nodes.

Managed expansion described in step 3 is achieved via Algorithm 15. The algorithm first checks whether there is only one interesting node and it should simply stop (lines 1 to 3). If there are more than one interesting node, it constructs $T_n$, used to keep track of all visited nodes, to initially contain all nucleus nodes (lines 4 and 5). The rest of the algorithm then incrementally expands all the nuclei until they are all connected (line 8 as $N_i$ - 1 edges are needed to connect $N_i$ nuclei) or when all nodes in the graph have been visited (line 26). In addition, we use variable *progress* (lines 7, 9 and 12) to keep track of the progress of graph expansion and stops algorithm if no progress has been made during the last iteration (line 8). A distance variable $d$ is used to manage the incremental expansion. As the expansion progresses, each of the encountered nodes is checked to see whether its distance

---

**Algorithm 15** Grow Interesting Subgraphs

---

**Input**: Global edge set $E_a$ in composition graph, global node set $N_a$ in composition graph, global edge references set $E_b$, global node reference set $N_b$, number of interesting node references $N_i$, and temporary node reference set $T_n$

**Variables**: distance $d$, set of paths $S_p$ connecting nucleus nodes, progress indication *progress*

```
 1: if Ni = 1 then
 2:     return;
 3: end if
 4: create Tn;
 5: add all nucleus nodes to Tn;
 6: d = 1;
 7: progress = true;
 8: while Sp.size < Ni − 1 and progress = true do
 9:     progress = false;
10:     for all nu ∈ Tn such that nu.distance = (d − 1) do
11:         for all e ∈ Eb such that e connects to nu do
12:             progress = true;
13:             identify the other node n that e connects to;
14:             if n.distance is not set then
15:                 n.distance = d;
16:                 n.nucleus = nu.nucleus;
17:                 Tn.add(n);
18:             else
19:                 // Found a potential merge point
20:                 if n.nucleus ≠ nu.nucleus then
21:                     if e ∉ Sp then
22:                         mark as connected the edge in Ea that e refers to;
23:                         connectPathToNucleus(Tn, nu);
24:                         connectPathToNucleus(Tn, n);
25:                         Sp.add(e);
26:                         if Sp.size() = Ni − 1 or Tn.size() = Eb.size() then
27:                             return;
28:                         end if
29:                     end if
30:                 end if
31:             end if
32:         end for
33:     end for
34:     d++;
35: end while
```

---

attribute is already set. If this is not set (line 14), then the node must be a newly encountered node and the algorithm sets the distance (line 15), tags it as belonging to the same nucleus group (line 16) and indicates that the node has been visited by adding it into $T_n$ (line 17). If the node is a previously visited node (line 18) and it is associated to a nucleus group different from the current one (line 20), then a merge point (see Figure 4.11) is potentially encountered. To be sure, the algorithm checks whether the edge extending to the node just encountered has already been visited (line 21). If not, it marks the corresponding edge in $E_a$ as connected and

---

**Algorithm 16** Connect Path to Nucleus

---

$connectPathToNucleus(T_n, n_1)$

**Input**: Global edge set $E_a$ in composition graph

**Variables**: Distance $d_1$, node $n_2$ and edge $e_{12}$

1:  $d_1 = n_1.distance$;
2:  **while** $d_1 > 0$ **do**
3:     **if** $n_1$ is a group node **then**
4:        mark as connected all interesting edges in $E_a$ referred to by nodes in the group;
5:     **end if**
6:     **for all** $n_2 \in T_n$ such that $n_2.d = d_1 - 1$ **do**
7:        find edge $e_{12}$ connecting $n_1$ and $n_2$;
8:        mark as connected the edge in $E_a$ that $e_{12}$ refers to;
9:        $n_1 = n_2$;
10:      break;
11:     **end for**
12:     $d_1--$;
13: **end while**

---

then invokes $connectPathToNucleus()$, indicates that the edge has been visited by adding it into $S_p$ (line 25), and checks whether the stop criteria (line 26) have been met. Algorithm 16 lists the algorithm used in $connectPathToNucleus()$ to mark all edges from the encountered node and leading to the corresponding nucleus node as connected. The traversal of a group node (line 3) would trigger additional expansion (line 4) that would mark all interesting edges in the corresponding group also as connected.

**Complexity Analysis of Graph Expansion Algorithms**

We now analyze the complexity of our algorithms (Algorithms 13, 14, 15 and 16) for establishing the fully connected graph. Table 4.5 lists relevant variables used in our complexity analysis.

In Algorithm 13, line 7 has a factor of $E \times R_e$. If we use $T_{coalesce}^{(i)}$ to denote the time it takes for the $i^{th}$ iteration of function $coalesce()$, then lines 8 through 32 have a factor of $E \times R_e \times (T_{coalesce}^{(1)} + N \times R_n \times C \times logE)$, where $T_{coalesce}^{(1)}$ can be calculated based on Algorithm 14 as:

$$T_{coalesce}^{(1)} = O[logN \times log(N \times R_n) + E \times R_e + C \times T_{coalesce}^{(2)}]$$
$$\approx O[C^d \times E \times R_e]$$

In fact, as $coalesce()$ is being called recursively the size of $S_e$ gets smaller and smaller. Thus, while we are rounding off less significant terms in the calculation for

**Table 4.5**  Symbols and Parameters

| *Variables* | |
|---|---|
| $N$ | Number of nodes in the network |
| $C$ | Average number of connections per node |
| $E$ | Number of edges in the network $= C \times N$ |
| $R_n$ | Rate of "interesting" nodes |
| $R_e$ | Rate of "interesting" edges |
| $d$ | Average depth of recursion for *coalesce*() |
| $T_{coalesce}^{(i)}$ | Time for $i^{th}$ iteration of *coalesce*() |
| $T_1$ | Time for coalescing interesting nodes |
| $T_{ctn}$ | Time for *connectPathToNucleus*()() |
| $T_2$ | Time for growing subgraph |
| $T$ | Total time |

$T_{coalesce}^{(1)}$, we are also being conservative in using a steady size for $S_e$ for simplicity. Consequently,

$$
\begin{aligned}
T_1 &= O[E \times R_e + E \times R_e \times (T_{coalesce}^{(1)} + N \times R_n \times C \times logE)] \\
&= O[C \times N \times R_e + C \times N \times R_e \times (T_{coalesce}^{(1)} + N \times R_n \times C \times log(C \times N))] \\
&\approx O[(C^{d+2} \times R_e^2 + C^2 \times R_e \times R_n \times logN) \times N^2]
\end{aligned}
$$

We now calculate $T_2$ for Algorithm 15. Starting with line 5, we see that it has a factor of $N \times R_n$. Line 8 will be executed on average a number of times equivalent to the expansion depth, which is $\frac{N}{N \times R_n}$. Line 10 has a factor of $N$. Line 11 has a factor of $C$. Line 20 has a factor of $logE$. Finally, $T_{ctn}$ from *connectPathToNucleus*() has a factor of $N$. Thus $T_2$ can be calculated using:

$$
\begin{aligned}
T_2 &= O[\frac{N}{N \times R_n} \times N \times R_n \times C \times logE \times N)] \\
&= O[N \times C \times log(C \times N) \times N] \\
&= O[C \times N^2 logN]
\end{aligned}
$$

As a result,

$$
T = T_1 + T_2 \approx O[(C^2 \times R_e \times R_n + C) \times N^2 logN + C^{d+2} \times R_e^2 \times N^2]
$$

## Complexity Simulation of Graph Expansion Algorithms

We conducted experiment to simulate the performance of our graph expansion algorithms. Since these algorithms are generic and independent of whether nodes represent entities, services or operations, we decided to configure our simulation to start with a network of $N$ generic nodes, each has an $R_n$ chance of being picked as an interesting node. We control the connection density by varying $C_{outbound}$ representing the average outbound connections a node has and which is statistically equivalent to $\frac{C}{2}$. Each edge also has an $R_e$ chance of being identified as interesting. Figure 4.12 shows the relationship between the time it takes to establish fully connected graphs for interesting nodes and the number of nodes that a graph network has. For legend purpose, we use $R_n/R_e$ to differentiate the results. We see that when the average number of outbound connections per node $C_{outbound}$ is small, the corresponding time the graph expansion takes is relatively small and the functional curves are all tightly bundled together. As $C_{outbound}$ increases, these curves tend to diverge more and the time the graph expansion takes also increases.

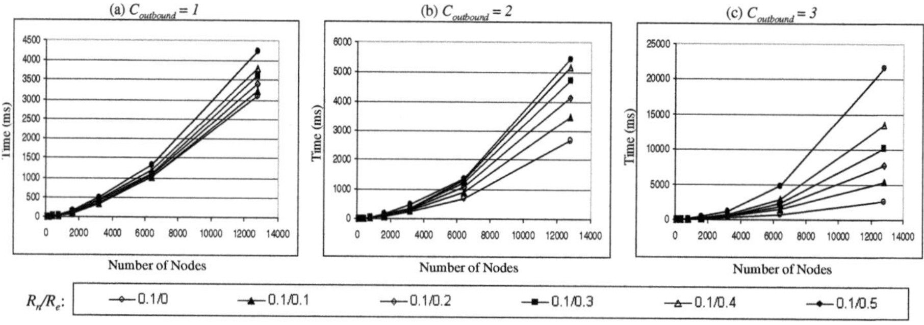


**Fig. 4.12**  Graph Expansion: Processing Time vs. Number of Nodes

Figure 4.13 shows the processing time as a function of the number of outbound connections $C_{outbound}$ for different values of $R_e$ and $N$. Note that when $C_{outbound} = 0$, there is no edges in the network. Consequently, the total processing time is practically 0.

### 4.3.3  Verification and Predictive Analysis via Runtime Simulation

Potentially composable Web services are identified in the screening phase based on ontologies, which contain static information about the semantics of those services. Such composability may not survive the dynamic runtime configuration that contains realistic external conditions and inter-dependencies among service oper-

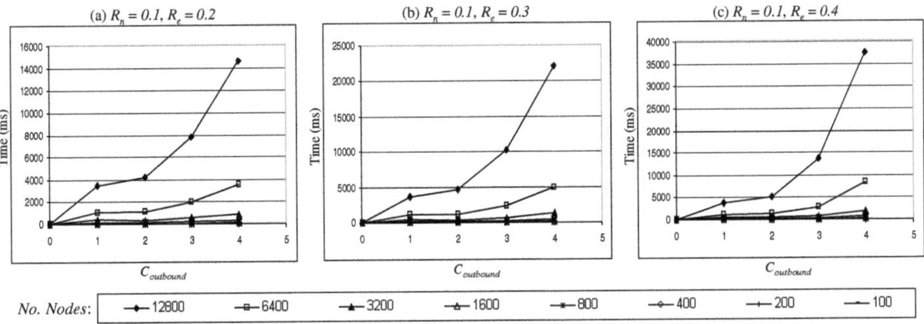



**Fig. 4.13**  Graph Expansion: Processing Time vs. Number of Outbound Connections per Node

ations. For example, a biological Web service may specify conditions necessary to enable or disable a biological process. These conditions may include temperature, parameter locale and quantity, kinetic energy, etc. Figure 4.14 illustrates that external conditions such as the surrounding temperature and light may play an important role in the establishment of a pathway. It also shows that the inter-dependencies among service operations can be due to parameter quantity, kinetic energy, accumulation and dissipation rates. For example, in order for pathway segment $EntityX : OperationA \rightarrow EntityY : OperationB$ to be established, the initial amount of substance $p$ generated by operation $A$ from entity $X$ should be greater than what is needed by operation $B$ from entity $Y$ or $p$'s accumulation due to operation $A$'s contribution exceeds its dissipation such that enough amount of $p$ needed by operation $B$ can be accumulated. In addition, for such a pathway to be established, the quantity and the kinetic energy of $p$, which might be boosted by the surrounding temperature or light before it dissipates may both need to be within a certain range.

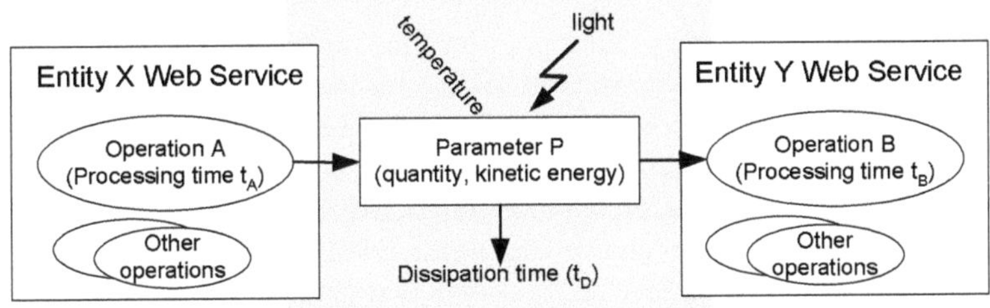



**Fig. 4.14**  Operation Recognition Dynamics

While it may not be feasible to account for the external conditions and operation inter-dependencies during the screening phase since they would force it to become too selective and computationally more expensive, they need to be considered in the evaluation phase to ensure that the composition leads identified in the screening phase really exist. The verification of composition validity can be carried out using a simulation environment, where functions of corresponding Web services can be invoked in the order as identified in composition flow networks. A composition lead identified in the screening phase indicates the potential possibility of a composition based on service and operation recognition. Verification aims at determining if segments of an identified composition flow network can indeed be enabled with a chain of relevant conditions.

The second important aspect of runtime simulation is its ability to support predictive analysis. Based on composition flow networks established from the screening phase and later highlighted in the interactive hypothesis formulation sub-phase, the user may attempt to conjecture a certain outcome from indirect relationships that are derived from the way the composition flow network is laid out. Section 5.4.0.9 outlines our simulation strategies that can be used to test out corresponding hypotheses in the case of pathway discovery.

### *4.3.4 Subjective Evaluation*

In addition to objective measures for both interestingness and usefulness, user evaluating these aspects of a service composition may choose to use subjective measures. The reference base of such measures may be personal knowledge, belief, bias and needs. Unfortunately, approaches based solely on subjective measures tend to inhibit us from getting interesting and useful compositions that were not thought of. An extreme case of relying on subjective measures to carry out Web service mining is the traditional composition approach where the user issues a query specifying the composition in pursuit to start the search process. Inspired by the drug discovery process, we rely on objective measures to first reduce the candidate population in an automatic fashion. More expensive and time consuming subjective evaluation can then be carried out in a much smaller pool for final determination.

We described our strategy in Section 4.3.2.1 for identifying and highlighting interesting segments within a composition network. These segments are presented to the user as visual aids for focusing on parts of the network that might lead to the identification of ultimately useful composition networks. Based on these visual aids and subjective interestingness and/or usefulness criteria, the user may then select a subset of nodes representing service, operation and parameters within the composition network for further exploration. We also presented algorithms in Section 4.3.2.2 for linking interesting segments and nodes into a fully connected graph, which can be presented to the user for hypothesis formulation. Simulation described in Section 4.3.3 can then be carried out to test out these hypotheses. Simulation results will be presented to the user, who can then correlate them with the hypotheses made earlier to see if such hypotheses can be either confirmed or rejected. Based on such analysis, the user may then make the ultimate determination as to whether the composition network under investigation is really useful. In Section 5.4.0.9, we analyze our simulation results from a real example.

# Chapter 5
# Application: Biological Pathway Discovery

In this chapter, we attempt to apply our service mining framework to the discovery of biological pathways. In Section 3.6, we introduced various entities inside a eukaryotic cell, as shown in Figure 3.7. These entities all have attributes and behaviors. The behaviors are manifested through entities' interactions with one another and the surrounding environment. Such interactions are the foundations for many of the pathways that are essential to the well being of our body. A biological pathway is a series of molecular interactions and reactions [18]. Biological pathways have traditionally been discovered manually based on experimental data such as gene expression data from microarrays, protein-protein interaction data from large-scale screening, and pathway data from previous discoveries. As a whole, biological pathways form the bridges that link much of the diverse range of biological data into a logical picture of why and how human genes and cells function the way they do. Disturbances and alterations in many of these pathways are also expected to be linked to various diseases.

The amount of data capturing biological entities and processes at various levels has recently increased manifold due to worldwide research projects in genomics, epigenomics and proteomics. Although these projects have begun answering many questions regarding how the human biological machinery works, much of the gold mine of biological information generated from these projects is still unexplored to a large extent: critical links hidden across various lab results are still waiting to be identified; isolated segments of potentially more comprehensive pathways are yet to be linked together. While early exposure of these hidden pathway linkages is expected to deepen our understanding of how diseases come about and help expedite drug discovery for treating them, it is now obvious that the complexity and enormity of information involved in the exposure of such hidden linkages may be too overwhelming for an unaided human mind to comprehend [55]. As a result, such exposure often requires the use of mining tools, which help elicit new knowledge and hypotheses by examining "a large corpus of old information" [91]. Unfortunately, approaches taken today for representing biological data focus on either pathway identification or pathway simulation, but not both. This consequently makes it diffi-

G. Zheng and A. Bouguettaya, *Web Service Mining: Application to Discoveries of Biological Pathways*, DOI 10.1007/978-1-4419-6539-4_5,

cult to devise effective mining tools. There are currently two major approaches used to represent biological entities: *free text based description* and *computer models*.

## 5.1 Free Text Based Description

Free text based approaches are mostly targeted at human comprehension. They use free text annotations and narratives [32, 30] to describe attributes and processes of biological entities and store them in various databases. These include GenBank [76], Database of Interacting Proteins (DIP) [94], Kyoto Encyclopedia of Genes and Genomes (KEGG) [63, 59], Swiss-Prot (a protein knowledgebase maintained by Swiss Institute for Bioinformatics [9]), and Cytokines Online Pathfinder Encyclopedia (COPE) [53]. A major disadvantage with these annotations and narratives is their lack of structure and interfaces. These are required for a computer application to "understand" the various concepts and often complex relationships among these concepts. Although several Natural Language Processing (NLP) approaches (e.g., [77, 104, 69, 87]) have been devised for the purpose of pathway identification, due to the limitation of static free text-based representation of biological data that these approaches target, they inherently lack the capability of verifying the validity and supporting 'what-if' analyses of identified pathways through computer-based simulation.

## 5.2 Computer Models

A computer model of a biological entity can be created based on lab discoveries and hypotheses. Such models can be both expressive (for human comprehension) and structured (for computer consumption) and thus provide a better alternative to free-text annotations. They can be understood by a human through their visual representations and by a computer through their constituent constructs. A major advantage of computer models is their readiness for execution with their inherent processes. By executing processes of computer models in a simulation, we expect to verify the validity of previously identified pathways linking real biological entities as represented by these models. When their processes are expressed as a function of surrounding conditions (e.g., availability of nutrients and energy), computer models would also have the inherent capability of responding to perturbations in these conditions, making it possible to study the effects of the perturbations on the pathways to the extent allowed by these models. Computer models have been pursued in [2], [3], [28], [31], [36], [93] and [61]. Worth particular mention among these models is the work done by Cardelli [31]. Cardelli took an information science perspective in modeling biological systems and processes. He categorized four large classes of *macromolecules* in a cell as the following chemical toolkits:

- *Nucleic acids*, including *RNA* as lists and *DNA* as doubly-linked lists.

- *Proteins*, whose surface features determine their functionalities. As data structures, proteins are modeled as records of features or objects in object-oriented programming sense.
- *Lipids*, among which the *membranes* were studied. As data structures, membranes are modeled as containers with an active surface that acts as an interface to its contents.
- *Carbohydrates*, among which *oligosaccharides* were studied. As data structures, oligosaccharides are models as trees.

These toolkits are then treated as abstract machines with states and operations. While limiting due to its fictional interpretations, these abstract machines are grounded in realistic biological systems and thus provides a certain degree of confidence in understanding the higher principles of biological organization. The *Gene Machine* (better known as Gene Regulatory Networks) performs information processing tasks within the cell. It regulates all other activities, including assembly and maintenance of the other machines, and the copying of itself. The *Protein Machine* (better known as Biochemical Networks) performs all mechanical and metabolic tasks, and also some signal processing. The *Membrane Machine* (better known as Transport Networks) separates different biochemical environments, and also operates dynamically to transport substances via complex, discrete, multi-step processes.

However elaborate they might be, these above models are usually constructed to simulate entities in an isolated local environment (as compared to the global Web environment), limited to the study of known pathways (e.g., cell death, growth factor activated kinase in BPS [2]), and lack the ability to facilitate the discovery of new pathways linking models that are independently developed.

## 5.3  A Service Oriented Modeling and Deployment Approach

We propose to model and deploy biological entities and their processes as Web services to bridge the gap between the above two representation approaches. Using this strategy, biological processes are modeled as Web service operations and exposed via standard Web service interfaces. An operation may consume some input substance meeting a set of preconditions and then produce some output substance as a result of its invocation. Some of these input and output substances may themselves carry processes that are known to us and thus can also be modeled and deployed as Web services. Domain ontologies containing definition of various entity types would be used by these Web services when referring to their operation inputs and outputs. This service oriented process modeling and deployment strategy opens up new interesting possibilities. First, like existing natural language processing approaches, it allows us to use service mining tools to proactively and systematically sift through Web service description documents in the service registry for automatic discovery of previously hidden pathways. Second, it brings about unprecedented opportunity for validating such pathways right on the Web through direct invocation of involved services. This second capability also makes it possible to carry out simulation-based predictive analyses of interactions involving a large number of entities modeled by these services. When enough details are captured in the service process models, this in-place invocation capability presents an inexpensive and accessible alternative to existing *in vitro* and/or *in vivo* exploratory mechanisms. In the presence of a large amount of biological information already made available in many formats from various sources, the adoption of this approach will undoubtedly incur an initial cost. However, this cost is only one-time and will be relatively trivial compared to the on-going development cost of various coupling mechanisms required between applications and limited potential such coupling mechanisms can offer.

We present in this section how we describe biological processes in WSDL/WSML and deploy them as Web services [110] using Web Service Modeling eXecution environment (WSMX). We then apply our service mining framework to Web service models of biological processes and present pathway networks discovered in our experiments. We start by describing the conceptual service oriented models of these processes.

### 5.3.1 Conceptual Service Oriented Models

We compiled a list of process models based on [7], [23], [45], [52], [64] and [105]. In addition to describing process models, these sources also reveal some simple relevant pathways that can be manually put together. We show examples of process models and corresponding simple pathways in Figure 5.1.

Multiple examples of *promotion, inhibition* and *indirect recognition* can be found in these simple pathways. For example, Figure 5.1 (a) shows that 15 LO provides an operation called *produce LXA4*, which promotes the service of LXA4, short for Lipoxin A4. Figure 5.1 (d) shows that upon injury, Leukotriene B4 (LTB4) recruits Neutrophil, promoting its service of producing COX2. Figure 5.1 (j) shows that Gastric Juice's service can inhibit the services of both Stomach Cell and Mucus. Examples of *indirect recognition* can be found in Figure 5.1 (i), where the service of Phospholipase A2 (PLA2) can liberate Arachidonic Acid, which can in turn be used as input to the *produce PGG2* operation of COX1's service or the *produce PGE2* operation of the COX2 service, where PGG2 and PGE2 stand for Prostaglandin G2 and Prostaglandin E2, respectively. Examples of pre- and post-conditions can be found in Figure 5.1 (h), where NF-$\kappa$B/Rel when not phosphorylated can translocate from cytoplasm to cell nucleus, where it can stimulate proinflammatory gene transcription. NF-$\kappa$B/Rel's service, however, may be inhibited by the I$\kappa$B service if NF-$\kappa$B/Rel is bound by its corresponding operation when I$\kappa$B is not phosphorylated. We use process models such as those in Figure 5.1 as references when developing WSDL and WSML Web services in Sections 5.3.3 and 5.3.4. We also use simple pathways manually constructed as references when we check the correctness of pathways automatically discovered in Section 5.4.0.8 using our mining algorithms.

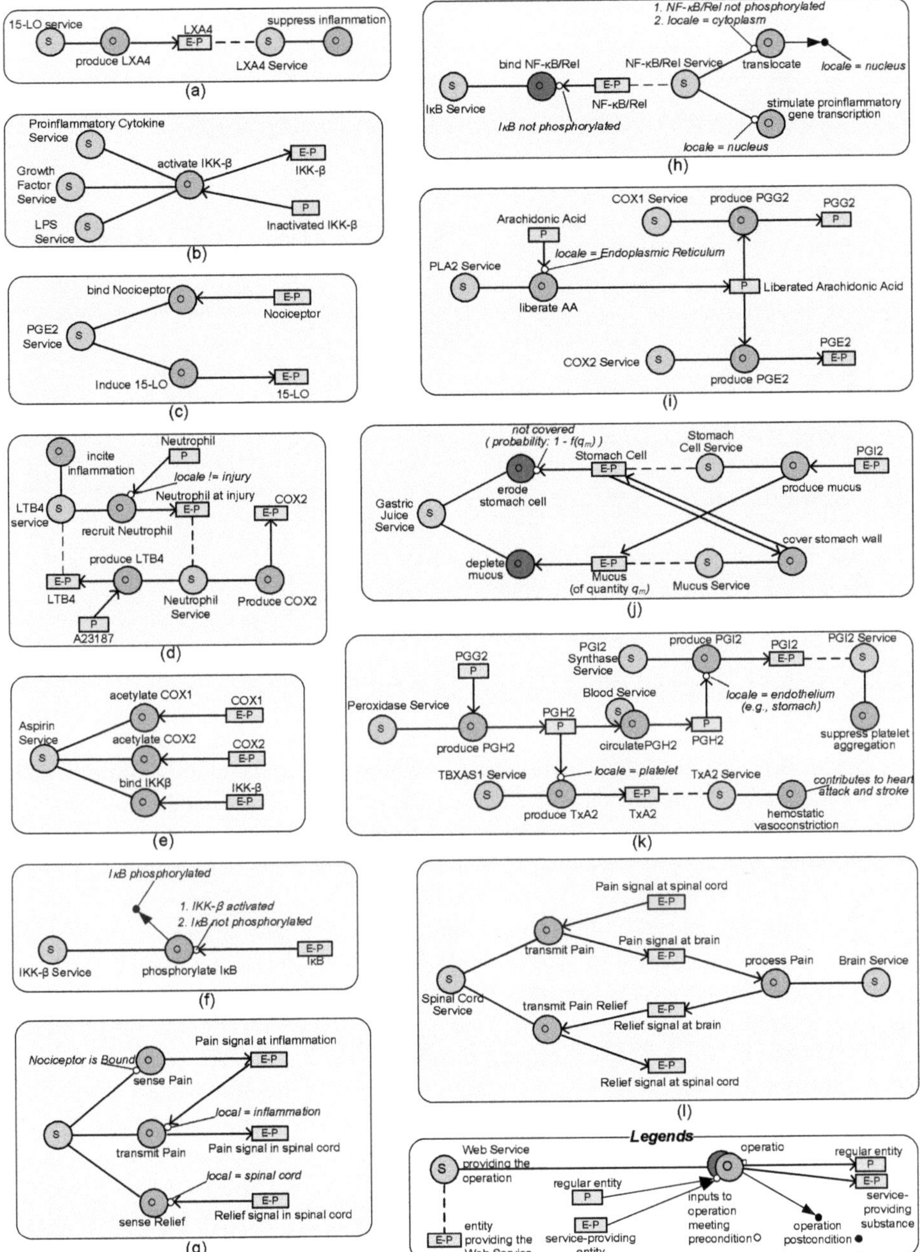

**Fig. 5.1**  Examples of Conceptual Process Model and Simple Pathway

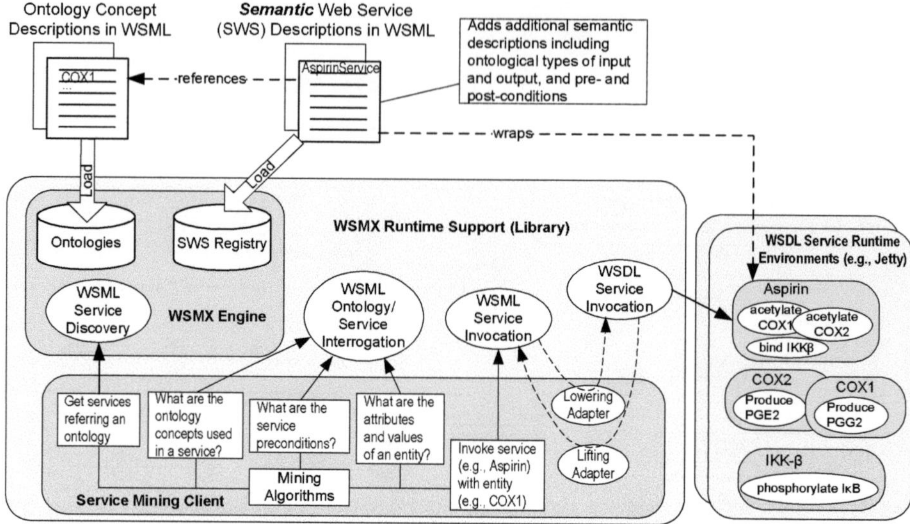


**Fig. 5.2**  Service Representation and Runtime Infrastructure

## 5.3.2  WSDL/WSML Description and Deployment

In this section, we present an overview of our service representation strategy and deployment environment. Figure 5.2 depicts the runtime infrastructure used to support the execution of various components. First, we model the actual process details for each type of biological entity as a WSDL [13] service and deploy these services using a Jetty Web server [6] (Section 5.3.3). Next, we expose the semantic interface (i.e., ontological types of operation input and output, pre- and post-conditions) of each WSDL service using Web Service Modeling Language (WSML) [10] as a Semantic Web service (Section 5.3.4). These WSML services are then deployed and discovered at runtime using the Web Services Modeling eXecution environment (WSMX) [14]. A WSMX engine is used to host the WSML services. It provides runtime interfaces to our service mining client, which aims at the discovery of interesting pathways linking relevant WSML services. Both lowering and lifting adapters (Section 5.3.5) are provided by the mining client to convert ontological entity instances used by WSML services to/from Simple Object Access Protocol (SOAP) messages used by WSDL services.

### *5.3.3 WSDL Service Modeling of Biological Processes*

We first define an XML schema (Figure 5.3) containing generic types such as *InputSubstance*, *OutputSubstance* and *BooleanResponse*. These types are intended to be used by biological process models to represent their input and output substances. For example, type *OutputSubstance* contains information about an output substance type, location and an generic boolean flag that can be used for passing additional information (e.g., the output of *activate IKK-β* in Figure 5.1 (a) would have this flag set as activated). In some cases, an operation may simply return a boolean response indicating whether the corresponding process represented by the operation has been invoked successfully. The XML schema is then run through an *xjc* [5] compiler to generate corresponding bean classes. We define each process model with a Java class based on these bean classes. Using Axis2 [1] running inside a Jetty Web server [6], these Java classes are exposed as WSDL Web services at runtime.

```xml
<?xml version="1.0"?>
<xsd:schema xmlns:xsd="http://www.w3.org/2001/XMLSchema"
  xmlns="http://servicemining.org/"
  targetNamespace="http://servicemining.org/">

  <xsd:element name="ArachidonicAcid" type="InputSubstance" />

  <xsd:complexType name="InputSubstance">
    <xsd:sequence>
      <xsd:element name="type" type="xsd:string" minOccurs="1" maxOccurs="1" />
      <xsd:element name="location" type="xsd:string" minOccurs="1" maxOccurs="1" />
      <xsd:element name="amount" type="xsd:float" minOccurs="1" maxOccurs="1" />
      <xsd:element name="flag" type="xsd:boolean" minOccurs="1" maxOccurs="1" />
    </xsd:sequence>
  </xsd:complexType>

  <xsd:complexType name="OutputSubstance">
    <xsd:sequence>
      <xsd:element name="type" type="xsd:string" minOccurs="1" maxOccurs="1" />
      <xsd:element name="location" type="xsd:string" minOccurs="1" maxOccurs="1" />
      <xsd:element name="amount" type="xsd:float" minOccurs="1" maxOccurs="1" />
      <xsd:element name="flag" type="xsd:boolean" minOccurs="1" maxOccurs="1" />
    </xsd:sequence>
  </xsd:complexType>

  <xsd:complexType name="BooleanResponse">
    <xsd:sequence>
      <xsd:element name="result" type="xsd:boolean" minOccurs="1" maxOccurs="1" />
    </xsd:sequence>
  </xsd:complexType>

</xsd:schema>
```

**Fig. 5.3**  Schema for WSDL Services

### *5.3.4 WSML Service Wrapping of WSDL Service*

Although the internal details of biological processes can be modeled as WSDL Web services, WSDL itself does not provide elaborate mechanism for expressing the pre- and post-conditions of service operations. WSDL also lacks the semantics needed to unambiguously describe data types used by operation input and output messages.

We choose WSML [10] among others (e.g., OWL-S [19], WSDL-S [16]) to fill this
gap due to the availability of WSMX, which supports the deployment of ontologies
and Web services described in WSML. We categorize biological entities within our
mining context into several ontologies. These include *Fatty Acid, Protein, Cell,* and
*Drug.* They would all refer to a *Common* ontology containing generic entity types
such as *Substance,* the root concept of all entity types. We use *UnknownSubstance*
as a placeholder for process inputs that are not fully described in the literature. We
also create a *Miscellaneous* ontology capturing definitions of entity types found in
the literature that don't seem to belong to any domain. For illustration purposes, we
show the common ontology in Figure 5.4 and the protein ontology in Figure 5.5.
Figure 5.5 (b) shows the WSMT [12] rendering of the protein ontology listed in
Figure 5.5 (a).

```
wsmlVariant _"http://www.wsmo.org/wsml/wsml-syntax/wsml-rule"

namespace {_"http://servicemining.org/Ontologies/CommonOntology#",
        dc       _"http://purl.org/dc/elements/1.1#",
        wsml _"http://www.wsmo.org/wsml/wsml-syntax#"}

ontology CommonOntology

/*
 * concepts
 */

concept Substance
        locale ofType _string
        quantity ofType _decimal

concept UnknownSubstance subConceptOf Substance
        localeInjured ofType _boolean

concept Bool
        result ofType _boolean

instance uks memberOf UnknownSubstance
        quantity hasValue 1
        locale hasValue "inflammation"
        localeInjured hasValue true

concept Signal
        locale ofType _string

instance bool memberOf Bool
```

**Fig. 5.4**  The Common Ontology of Entity Types

Using these ontologies, we then wrap the semantic interfaces of existing WSDL
services as WSML services. WSML supports the descriptions of pre- and post-
conditions in the capability section and the ontological type description in the interface section. Figure 5.6 gives an example of each for the *NF_kappaB_Rel* service. The capability section states for the precondition that the input entity instance
named *nfkbr* should be of type *NF_kappaB_Rel* (defined in the protein ontology). In
addition, *nfkbr*'s locale should be cytoplasm and it should not be phosphorylated.
The interface section states that input entity *NF_kappaB_Rel* has grounding with the

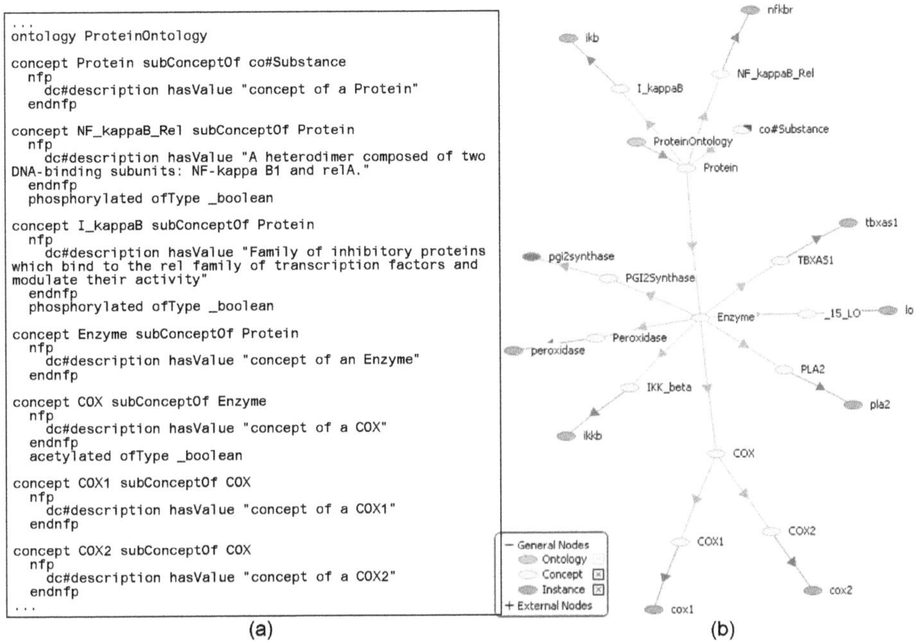

```
...
ontology ProteinOntology

concept Protein subConceptOf co#Substance
  nfp
    dc#description hasValue "concept of a Protein"
  endnfp

concept NF_kappaB_Rel subConceptOf Protein
  nfp
    dc#description hasValue "A heterodimer composed of two
DNA-binding subunits: NF-kappa B1 and relA."
  endnfp
  phosphorylated ofType _boolean

concept I_kappaB subConceptOf Protein
  nfp
    dc#description hasValue "Family of inhibitory proteins
which bind to the rel family of transcription factors and
modulate their activity"
  endnfp
  phosphorylated ofType _boolean

concept Enzyme subConceptOf Protein
  nfp
    dc#description hasValue "concept of an Enzyme"
  endnfp

concept COX subConceptOf Enzyme
  nfp
    dc#description hasValue "concept of a COX"
  endnfp
  acetylated ofType _boolean

concept COX1 subConceptOf COX
  nfp
    dc#description hasValue "concept of a COX1"
  endnfp

concept COX2 subConceptOf COX
  nfp
    dc#description hasValue "concept of a COX2"
  endnfp
...
```



(a)                                                     (b)

**Fig. 5.5**  The Protein Ontology

input parameter of operation *translocate* of the corresponding WSDL service. The output from the WSDL service operation should be mapped to *NF_kappaB_Rel* as defined in the protein ontology.

To work with WSML, we have made slight adaptations to our screening algorithms so they can be applied directly to WSML services. First, we add a *provider* property in the non functional properties (nfp) section of each WSML service to indicate the corresponding ontological type of an entity that can provide the service. We use this information in our algorithms to establish the relationship between a service providing entity and the service it provides. Second, we add a *modelSource* property in the *nfp* section to indicate the source information that the model is based on. This information allows our algorithms to automatically identify interesting pathway segments within a discovered pathway network. Third, we add a *providerConsumable* property in the *nfp* section to indicate whether the service providing entity should be consumed along the invocation of its operation. For example, in order for mucus (Figure 5.1 (j)) to cover the wall of the stomach, the mucus itself will have to be consumed. Finally, our validation algorithm has been customized to work with the service interrogation Application Programming Interfaces (APIs) of the WSMX runtime library (Figure 5.2) for determining the overlap between the postcondition of a source operation and the precondition of a target operation. Unfortunately, WSML allows for the specification of pre- and post-conditions for only an entire service, but not its individual operations. Thus we have to split services that each originally has multiple operations into several services (e.g., *NF_kappaB_Rel_1_Service* and *NF_kappaB_Rel_2_Service*) so that different conditions can be individually specified for these operations. We use the name of these services to keep track of their relationship and use that information to merge these services towards the end of the screening phase.

```
wsmlVariant _"http://www.wsmo.org/wsml/wsml-syntax/wsml-rule"
namespace { _"http://servicemining.org/SWSs/NF_kappaB_Rel_1_Service#",
     po _"http://servicemining.org/Ontologies/ProteinOntology#",
     dc _"http://purl.org/dc/elements/1.1#",
     wsml _"http://www.wsmo.org/wsml/wsml-syntax#" }


webService NF_kappaB_Rel_1_Service
       nfp
               dc#contributor hasValue "George Zheng"
               _"http://owner" hasValue _"http://ServiceMining"
               _"http://modelSource" hasValue _"http://ServiceMining/h"
               _"http://provider" hasValue _"http://servicemining.org/Ontologies/
ProteinOntology#NF_kappaB_Rel"
               _"http://providerConsumable" hasValue _"http://servicemining.org/true"
       endnfp

   importsOntology
   {
           po#ProteinOntology
   }

capability translocate

   precondition
       definedBy
             ?nfkbr memberOf NF_kappaB_Rel[
             locale hasValue ?l,
             phosphorylated hasValue ?p] and
             (?l = "cytoplasm") and
             (?p = false).

   postcondition
       definedBy
             ?nfkbr memberOf NF_kappaB_Rel[
             locale hasValue ?l] and
             (?l = "nucleus").

interface NF_kappaB_Rel_1_ServiceInterface

     choreography NF_kappaB_Rel_1_ServiceChoreography
   stateSignature NF_kappaB_Rel_1_ServiceStatesignature

   importsOntology
   {
           po#ProteinOntology
   }

       in
             concept po#NF_kappaB_Rel withGrounding _"http://servicemining.org:8001/
NFkappaBRel?wsdl#wsdl.interfaceMessageReference(NFkappaBRel/translocate/in0)"

       out
             concept po#NF_kappaB_Rel

   transitionRules NF_kappaB_Rel_1_ServiceTransitionRules
```

**Fig. 5.6**   Semantic Interface Description in WSML

### 5.3.5 *Lowering and Lifting Adapters*

```
...

ontology AdapterOntology

concept xml2wsmlmapping
     instanceMappings impliesType  (1 *) _string
     valueMappings impliesType  (1 *) _string
     conceptOutput impliesType  (1 1) _string
     inputMessage impliesType  (1 1) _string

instance acetylateCOX2Response memberOf xml2wsmlmapping
     valueMappings hasValue { "//result=result"}
     conceptOutput hasValue "CommonOntology#Bool"
     inputMessage hasValue "//ns1:BOOL"

instance activateIKKBetaResponse memberOf xml2wsmlmapping
     valueMappings hasValue { "//location=locale(_string)", "//amount=quantity",
"//flag=activated"}
     conceptOutput hasValue "ProteinOntology#IKK_beta"
     inputMessage hasValue "//ns1:IKKBeta"

instance bindIKKBetaRelResponse memberOf xml2wsmlmapping
     valueMappings hasValue { "//result=result"}
     conceptOutput hasValue "CommonOntology#Bool"
     inputMessage hasValue "//ns1:BOOL"

instance bindNociceptorResponse memberOf xml2wsmlmapping
     valueMappings hasValue { "//location=locale(_string)", "//amount=quantity",
"//flag=isBound"}
     conceptOutput hasValue "NervousSystemOntology#Nociceptor"
     inputMessage hasValue "//ns1:Nociceptor"

instance bindNFkappaBRelResponse memberOf xml2wsmlmapping
     valueMappings hasValue { "//result=result"}
     conceptOutput hasValue "CommonOntology#Bool"
     inputMessage hasValue "//ns1:BOOL"

instance circulatePGH2Response memberOf xml2wsmlmapping
     valueMappings hasValue { "//location=locale(_string)", "//amount=quantity"}
     conceptOutput hasValue "FattyAcidOntology#PGH2"
     inputMessage hasValue "//ns1:ProstaglandinH2"

instance coverStomachWallResponse memberOf xml2wsmlmapping
     valueMappings hasValue { "//location=locale(_string)", "//amount=quantity",
"//flag=coveredByMucus"}
     conceptOutput hasValue "CellOntology#Stomach_Cell"
     inputMessage hasValue "//ns1:StomachCell"
...

instance liberateAAResponse memberOf xml2wsmlmapping
     valueMappings hasValue { "//location=locale(_string)", "//amount=quantity",
"//flag=liberated"}
     conceptOutput hasValue "FattyAcidOntology#Arachidonic_Acid"
     inputMessage hasValue "//ns1:AA"
...

instance producePGG2Response memberOf xml2wsmlmapping
     valueMappings hasValue { "//location=locale(_string)", "//amount=quantity"}
     conceptOutput hasValue "FattyAcidOntology#PGG2"
     inputMessage hasValue "//ns1:ProstaglandinG2"
...
```

**Fig. 5.7**  Adapter Ontology for Lifting Adapter

When a WSML service is to be invoked, a *lowering adapter* (right end of service
mining client at the bottom of Figure 5.2) is used to parse out the attribute values
of the input entity and package them into a SOAP message, which is in turn used to
invoke the corresponding WSDL service. Upon receipt of the return SOAP message
from the WSDL service after its invocation, a *lifting adapter* is used to parse from

it attribute values that are subsequently used to create an instance of an ontological entity type as specified in an adapter ontology. We show in Figure 5.7 the adapter ontology used to provide the mappings for such conversion. As an example, the adapter ontology states that the *AA* field in the return SOAP message of the *liberateAA* operation of the corresponding WSDL service should be mapped to concept *Arachidonic_Acid* defined in the fatty acid ontology. In addition, the *location* field should be mapped to the *locale* attribute, the *amount* field should be mapped to the *quantity* attribute, and the *flag* field should be mapped to the *activated* attribute. Using this mechanism, an entity instance (e.g., liberated arachidonic acid in Figure 5.1 (i)) created after the invocation of a WSML service operation (e.g., *liberateArachidonicAcid* of the PLA2 Service) can be instantiated accordingly and in turn used as an input to another WSML service operation (e.g., *producePGG2* of the COX1 Service).

## 5.4 Experiment

In this section, we describe an end-to-end example of how we apply our service mining framework to the discovery of pathways linking service-oriented biological processes. Figure 5.8 shows our testbed infrastructure. Using this testbed, the development of our experiment involves eight steps: *WSDL service development, Ontology development, WSML service development, WSML service invocation, Scope specification, Focused library generation, Screening*, and *Evaluation*. We describe in detail each of these steps in the following sections.

### 5.4.0.1 WSDL Service Development

A collection of WSDL Web services are developed and deployed using a Jetty Web server as described in Section 5.3.3. For the experiment, the Jetty Web server runs on a Windows XP machine registered as *servicemining.org* and exposes all the WSDL services through port *8001*. Thus, a request for the *COX1* WSDL service through *http://servicemining.org:8001/COX1?wsdl* returns the content of the service as shown in Figure 5.9. We show in Figure 5.10 the Java class used by Axis2 [1] running inside the Jetty Web server to generate the WSDL service for COX1. Note that both *InputSubstance* and *OutputSubstance* used for the type of operation input and output parameters are defined in Figure 5.3. Along with the *COX1* service, 24 other WSDL services are also deployed in the same fashion. Table 5.1 lists all the WSDL services used in our experiment with their operations, corresponding input and output parameters, and a simple description of logic used in the operation. We use *default* to denote default operation logic that can be derived using the following rules:

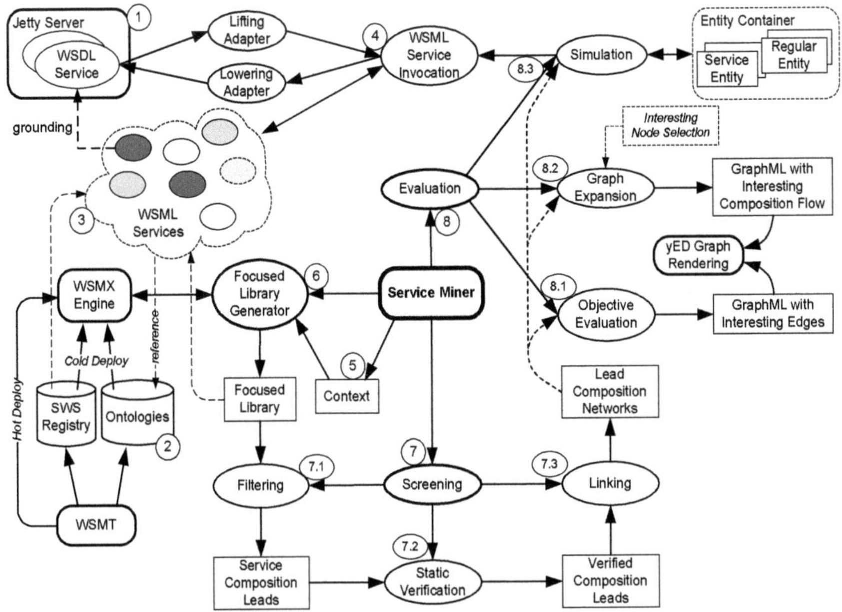

**Fig. 5.8** Testbed Infrastructure

```xml
<?xml version="1.0" encoding="UTF-8"?><definitions xmlns="http://schemas.xmlsoap.org/wsdl/"
xmlns:tns="http://servicemining.org/" xmlns:xsd="http://www.w3.org/2001/XMLSchema" xmlns:soap="http://
schemas.xmlsoap.org/wsdl/soap/" targetNamespace="http://servicemining.org/" name="COX1Service">
  <types>
    <xsd:schema>
      <xsd:import schemaLocation="http://servicemining.org:8001/COX1?xsd=1" namespace="http://
servicemining.org/"></xsd:import>
    </xsd:schema>
  </types>
  <message name="producePGG2">
    <part element="tns:producePGG2" name="parameters"></part>
  </message>
  <message name="producePGG2Response">
    <part element="tns:producePGG2Response" name="parameters"></part>
  </message>
  <portType name="COX1">
    <operation name="producePGG2">
      <input message="tns:producePGG2"></input>
      <output message="tns:producePGG2Response"></output>
    </operation>
  </portType>
  <binding name="COX1PortBinding" type="tns:COX1">
    <soap:binding style="document" transport="http://schemas.xmlsoap.org/soap/http"></soap:binding>
    <operation name="producePGG2">
      <soap:operation soapAction="producePGG2"></soap:operation>
      <input>
        <soap:body use="literal"></soap:body>
      </input>
      <output>
        <soap:body use="literal"></soap:body>
      </output>
    </operation>
  </binding>
  <service name="COX1Service">
    <port name="COX1Port" binding="tns:COX1PortBinding">
      <soap:address location="http://servicemining.org:8001/COX1"></soap:address>
    </port>
  </service>
</definitions>
```

**Fig. 5.9** COX1 WSDL Service

```
@WebService(name = "COX1", targetNamespace = "http://servicemining.org/")
@SOAPBinding(style = Style.DOCUMENT, use = Use.LITERAL, parameterStyle = ParameterStyle.WRAPPED)
public class COX1 {

    @WebMethod(operationName="producePGG2", action="producePGG2")
    @WebResult(name="ProstaglandinG2", partName="OutputSubstance", targetNamespace = "http://
servicemining.org/")
    public OutputSubstance producePGG2(
        @WebParam(name="ArachidonicAcid", partName="InputSubstance", targetNamespace = "http://
servicemining.org/")
        InputSubstance is) {

        System.out.println("producePGG2 operation of COX1 service invoked with: type=" + is.getType()
+ " location=" + is.getLocation() + " amount=" + is.getAmount() + " liberated=" + is.isFlag());

        ObjectFactory of = new ObjectFactory();

        OutputSubstance os = of.createOutputSubstance();

        if (is.getType().equals("ArachidonicAcid")) {
            os.setType("ProstaglandinG2");
            os.setLocation("platelet");
            os.setAmount(is.getAmount());
        }

        return os;
    }
}
```

**Fig. 5.10**  Java Class Used to Generate COX1 WSDL Service

- If the returned object is to be of type *OutputSubstance*, then the default logic is to set the *location* and *amount* attributes to be equal to those used for in the input object.
- If the return object is to be of type *BooleanResponse*, then the default logic is to check whether the *type* attribute of the input object is the same as expected. If so, set the *result* attribute of *BooleanResponse* to *true*. Otherwise, set it to *false*.

We list for each operation in the third column of Table 5.1 the expected value for the *type* attribute.

### 5.4.0.2 Ontology Development

A collection of ontologies in WSML are developed as described in Section 5.3.4 and validated using the WSMX engine. Figure 5.11 shows ontologies used in our experiment. These include: *CommonOntology, DrugOntology, ProteinOntology, NervousSystemOntology, FattyAcidOntology, CellOntology,* and *MiscOntology*. Note the inheritance (i.e., *subConceptOf* relationship between various concepts as graphically depicted with an arrow. To facilitate instance discovery, we also instantiated entities for many of the concepts. These are labeled with lower case names and are always at the leaf node positions. Table 5.2 shows ontological concepts that have attributes not included in the WSMT rendering. Since many of the ontological concepts are *subConceptOf Substance* in the *CommonOntology*, they all inherit the *locale* and *quantity* attribute from *Substance*. For our experiment, we place all the ontology WSML files under a directory on the same machine hosting the WSMX engine. We specify the location of the ontology folder with the *wsmx.resourcemanager.ontologies* variable in the *config.properties* file for the WSMX. Upon startup, WSMX uses the above variable to find these ontologies and load them into the runtime environment. During startup, the WSMX checks

**Table 5.1** Descriptions of WSDL Services

| Service | Operations | InputSubstance Attribute *type* | OutputSubstance Attribute *type* or BooleanResponse | Operation Logic |
|---|---|---|---|---|
| _15_LO | *produceLAX4* | UnknownStuff | LXA4 | set output location to inflammation |
| Aspirin | *acetylateCOX1* | COX1 | BooleanResponse | default |
| | *acetylateCOX2* | COX2 | BooleanResponse | default |
| | *bindIKK_beta* | IKKBeta | BooleanResponse | default |
| Blood | *circulatePGH2* | ProstaglandinH2 | ProstaglandinH2 | set output location to endothelium |
| Brain | *processPain* | Pain | Relief | set output location to brain |
| COX1 | *producePGG2* | ArachidonicAcid | ProstaglandinG2 | set output location to platelet |
| COX2 | *producePGE2* | ArachidonicAcid | ProstaglandinE2 | default |
| GastricJuice | *erodeStomachCell* | StomachCell | StomachCell | If input is covered by mucus, flag output as not covered. Otherwise, set the amount of output to 0 |
| | *depleteMucus* | Mucus | BooleanResponse | default |
| IkappaB | *bindNFkappaBRel* | NFkappaBRel | BooleanResponse | default |
| IKK_beta | *phosphorylateIkappaB* | IkappaB | BooleanResponse | default |
| LPS | *activateIKKBeta* | IKKBeta | IKKBeta | set output as activated |
| LTB4 | *recruitNeutrophil* | Neutrophil | Neutrophil | flag locale of output as injured |
| | *inciteInflammation* | UnknownStuff | BooleanResponse | default |
| LXA4 | *suppressInflammation* | UnknownStuff | BooleanResponse | default |
| Mucus | *coverStomachWall* | StomachCell | StomachCell | flag output as covered |
| Neutrophil | *produceLTB4* | A23187 | LTB4 | default |
| | *produceCOX2* | UnknownStuff | COX2 | flag output as not acetylated |
| NFkappaBRel | *translocate* | NFkappaBRel | NFkappaBRel | set output location to nucleus |
| | *stimulatePGTranscription* | NFkappaBRel | BooleanResponse | default |
| Nociceptor | *sensePain* | Nociceptor | Pain | set output location to inflammation |
| | *transmitPain* | Pain | Pain | set output location to spinal cord |
| | *senseRelief* | Relief | BooleanResponse | default |
| Peroxidase | *producePGH2* | ProstaglandinG2 | ProstaglandinH2 | default |
| PGE2 | *induce15LO* | UnknownStuff | _15_LO | set the amount of output to 10% of that of input |
| | *bindNociceptor* | Nociceptor | Nociceptor | flag output as bound |
| PGI2 | *suppressPlateletAggregation* | UnknownStuff | BooleanResponse | default |
| PGI2Synthase | *producePGI2* | ProstaglandinH2 | ProstaglandinI2 | default |
| PLA2 | *liberateArachidonicAcid* | AA | AA | flag output as liberated |
| SpinalCord | *transmitPain* | Pain | Pain | set output location to brain |
| | *transmitPainRelief* | Relief | Relief | set output location to spinal cord |
| StomachCell | *produceMucus* | ProstaglandinI2 | Mucus | default |
| TBXAS1 | *produceTxA2* | ProstaglandinH2 | TxA2 | default |
| TxA2 | *vasoconstriction* | UnknownStuff | BooleanResponse | default |

whether there are problems within these ontologies. WSMX ontologies can also be loaded (or hot deployed as depicted in Figure 5.8) into a running instance of WSMX through WSMT. However, we did not use that feature in our experiment.

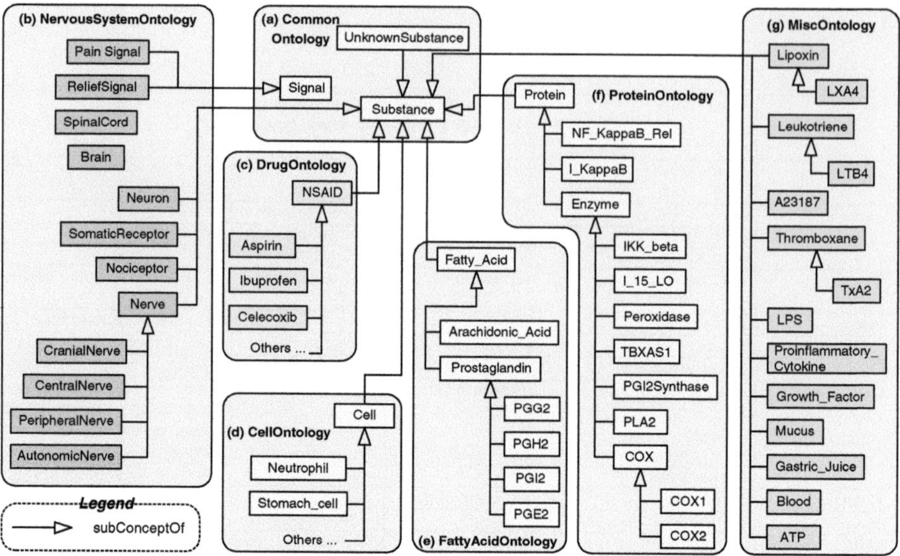



**Fig. 5.11**　Ontologies Used in Experiment

**Table 5.2**　Attributes of Ontological Concepts

| Ontology | Concept | Attribute | Attribute Type |
|---|---|---|---|
| CommonOntology | Substance | locale | _string |
| | | quantity | _decimal |
| | UnknownSubstance | localeInjured | _boolean |
| | Bool | result | _boolean |
| | Signal | locale | _string |
| ProteinOntology | NF_kappaB_Rel | phosphorylated | _boolean |
| | I_kappaB | phosphorylated | _boolean |
| | COX | acetylated | _boolean |
| | IKK_beta | activated | _boolean |
| NervousSystemOntology | Nociceptor | isBound | _boolean |
| FattyAcidOntology | Arachidonic_Acid | liberated | _boolean |
| | | localeInjured | _boolean |
| MiscOntology | A23187 | localeInjured | _boolean |

### 5.4.0.3　WSML Service Development

A collection of WSML services are developed as described in Section 5.3.4, where an example of WSML service is given in Figure 5.6. As indicated in Figure 5.8, WSML services relies on ontologies to define the type of their operation input and output parameters. Table 5.3 lists all the WSML services used in our experiment with the following information:

**Table 5.3**  Descriptions of WSML Services

| Service | Nfp | Capability | Precondition | Postcondition | In | Out |
|---|---|---|---|---|---|---|
| _15_LOService | a/false | *productLAX4* | default | default | UnknownSubstance | LXA4 |
| Aspirin_1_Service | e/false | *acetylateCOX1* | ?cox1 memberOf COX1[ acetylated hasValue ?a] and (?a = false). | default | COX1 | Bool |
| Aspirin_2_Service | e/false | *acetylateCOX2* | ?cox2 memberOf COX2[ acetylated hasValue ?a] and (?a = false). | default | COX2 | Bool |
| Aspirin_3_Service | e/false | *bindIKK_beta* | default | default | IKK_beta | Bool |
| BloodService | k/false | *circulatePGH2* | default | default | PGH2 | PGH2 |
| BrainService | l/false | *processPain* | ?ps memberOf PainSignal[locale hasValue ?l] and (?l = "brain"). | ?rs memberOf ReliefSignal[locale hasValue ?l] and (?l = "brain"). | PainSignal | ReliefSignal |
| COX1Service | i/false | *producePGG2* | ?aa memberOf Arachidonic_Acid [liberated hasValue ?l] and (?l = true). | default | Arachidonic_Acid | PGG2 |
| COX2Service | i/false | *producePGE2* | ?aa memberOf Arachidonic_Acid [liberated hasValue ?l] and (?l = true). | default | Arachidonic_Acid | PGE2 |
| GastricJuice_1_Service | j/false | *erodeStomachCell* | default | default | Stomach_Cell | Stomach_Cell |
| GastricJuice_2_Service | j/false | *depleteMucus* | default | default | Mucus | Mucus |
| I_kappaBService | h/true | *bindNF_kappaBRel* | ?nfkb memberOf NF_kappaB_Rel[ phosphorylated hasValue ?p] and (?p = false). | default | NF_kappaB_Rel | Bool |
| IKK_betaService | f/true | *phosphorylateI_kappaB* | ?ikb memberOf I_kappaB[ phosphorylated hasValue ?p] and (?p = false). | default | I_kappaB | Bool |
| LPSService | b/false | *activateIKK_Beta* | ?ikkbeta memberOf IKK_beta[ activated hasValue ?a] and (?a = false). | ?ikkbeta memberOf IKK_beta[ activated hasValue ?a] and (?a = true). | IKK_beta | IKK_beta |
| LTB4_1_Service | d/false | *recruit Neutrophil* | ?neutrophil memberOf Neutrophil[ localeInjured hasValue ?l] and (?l = false). | ?neutrophil memberOf Neutrophil[ localeInjured hasValue ?l] and (?l = true). | Neutrophil | Neutrophil |
| LTB4_2_Service | d/false | *inciteInflammation* | default | default | UnknownSubstance | Bool |
| LXA4Service | a/false | *suppressInflammation* | ?us memberOf UnknownSubstance[ locale hasValue ?l] and (?l = "inflammation"). | default | UnknownSubstance | Bool |
| MucusService | j/true | *coverStomachWall* | ?sc memberOf Stomach_Cell[ coveredByMucus hasValue ?c] and (?c = false). | ?sc memberOf Stomach_Cell[ coveredByMucus hasValue ?c] and (?c = true). | Stomach_Cell | Stomach_Cell |
| Neutrophil_1_Service | d/false | *produceLTB4* | ?neutrophil memberOf Neutrophil[ localeInjured hasValue ?l] and (?l = true). | default | A23187 | LTB4 |
| Neutrophil_2_Service | d/false | *produceCOX2* | ?uks memberOf UnknownSubstance[ localeInjured hasValue ?l] and (?l = true). | default | UnknownSubstance | COX2 |
| NF_kappaB_Rel_1_Service | h/true | *translocate* | ?nfkbr memberOf NF_kappaB_Rel[ locale hasValue ?l, phosphorylated hasValue ?p] and (?l = "cytoplasm") and (?p = false). | ?nfkbr memberOf NF_kappaB_Rel[ locale hasValue ?l] and (?l = "nucleus"). | NF_kappaB_Rel | NF_kappaB_Rel |
| NF_kappaB_Rel_2_Service | h/false | *stimulateProinflammatoryGeneTranscription* | ?nfkbr memberOf NF_kappaB_Rel[ locale hasValue ?l] and (?l = "nucleus"). | default | NF_kappaB_Rel | Bool |
| Nociceptor_1_Service | g/false | *sensePain* | ?nccpt memberOf Nociceptor[ isBound hasValue ?b] and (?b = true). | default | Nociceptor | PainSignal |
| Nociceptor_2_Service | g/false | *transmitPain* | ?ps memberOf PainSignal[ locale hasValue ?l] and (?l = "inflammation"). | ?ps memberOf PainSignal[locale hasValue ?l] and (?l = "spinal cord"). | PainSignal | PainSignal |
| Nociceptor_3_Service | g/false | *senseRelief* | ?ps memberOf ReliefSignal[ locale hasValue ?l] and (?l = "spinal cord"). | default | ReliefSignal | Bool |
| PeroxidaseService | k/false | *producePGH2* | default | default | PGG2 | PGH2 |
| PGE2_1_Service | c/false | *induce_15_LO* | default | default | UnknownSubstance | _15_LO |
| PGE2_2_Service | c/true | *bindNociceptor* | ?nccpt memberOf Nociceptor[ isBound hasValue ?b] and (?b = false). | ?nccpt memberOf Nociceptor [isBound hasValue ?b] and (?b = true). | Nociceptor | Nociceptor |
| PGI2Service | k/false | *suppressPlateletAggregation* | default | default | UnknownSubstance | Bool |
| PGI2Synthase Service | k/false | *producePGI2* | ?pgh2 memberOf PGH2[ locale hasValue ?l] and (?l = "endothelium"). | default | PGH2 | PGI2 |
| PLA2Service | i/false | *liberateArachidonicAcid* | ?aa memberOf Arachidonic_Acid[ liberated hasValue ?li, locale hasValue ?lo] and (?li = false) and (?lo = "endoplasmic_reticulum"). | ?aa memberOf Arachidonic_Acid[ liberated hasValue ?l] and (?l = true). | Arachidonic_Acid | Arachidonic_Acid |
| SpinalCord_1_Service | l/false | *transmitPain* | ?ps memberOf PainSignal[ locale hasValue ?l] and (?l = "spinal cord"). | ?ps memberOf PainSignal[ locale hasValue ?l] and (?l = "brain"). | PainSignal | PainSignal |
| SpinalCord_2_Service | l/false | *transmitPainRelief* | ?rs memberOf ReliefSignal[ locale hasValue ?l] and (?l = "brain"). | ?rs memberOf ReliefSignal[locale hasValue ?l] and (?l = "spinal cord"). | ReliefSignal | ReliefSignal |
| StomachCell Service | j/false | *produceMucus* | default | default | PGI2 | Mucus |
| TBXAS1Service | k/false | *produceTxA2* | ?pgh2 memberOf PGH2[ locale hasValue ?l] and (?l = "platelet"). | default | PGH2 | TxA2 |
| TxA2Service | k/false | *vasoconstriction* | default | default | UnknownSubstance | Bool |

- Service name. Note that in order to address the limitation of WSML that allows the specification of pre- and post-conditions for only the whole service, we break some of the services that have multiple operations so that pre- and post-conditions can be individually specified. These service are later merged together based on their names.
- *nfp* information in the form of *x/y* where *x* is part of the the *modelSource* property in the *nfp* section in the form of *http://ServiceMining/x* and corresponds to the index in Figure 5.1), and *y* is a boolean flag in the *nfp* section in the form of *http://servicemining.org/y* indicating whether the service provider is consumable during operation invocation. For instance, Table 5.3 shows that *NF_kappaB_Rel*, the provider for its corresponding service is consumable for the *translocate* operation but not the *stimulateProinflammatoryGeneTranscription* operation.
- Capability name, i.e., operation name
- Precondition and postcondition expressed in WSMO axiom logical expressions [11]. For instance, the precondition of the *translocate* operation of *NF_kappaB_Rel_1_Service* states that the corresponding parameter used during invocation is of type *NF_kappaB_Rel*. Its attribute *locale* is equal to *cytoplasm* and its attribute *phosphorylated* is equal to *false*. Similarly, the postcondition of the same operation states that the *locale* attribute has been changed to *nucleus*. The *default* for the two conditions listed for some of the services simply checks whether the parameter used during invocation is of the prescribed type.
- Ontological type for input and output parameters.

For our experiment, we place all the WSML files for services under a directory on the same machine hosting the WSMX engine. We specify the location of the ontology folder with the *wsmx.resourcemanager.webservices* variable in the *config.properties* file for the WSMX. Upon startup, WSMX uses the above variable to find these services and load them into the runtime environment. During startup, WSMX checks whether there are problems within these WSML services. WSML services can also be loaded (or hot deployed as depicted in Figure 5.8) into a running instance of WSMX through WSMT. However, we did not use that feature in our experiment.

### 5.4.0.4 WSML Service Invocation

During the development of the WSML services, we took an incremental approach by individually invoking each new pair of WSML and WSDL services to ensure that they have been developed properly. Such invocation is realized with the help of both lowering and lifting adapters (Section 5.3.5). For example, after we have developed and deployed *COX1Service*, we use a test program to first instantiate from *FattyAcidOntology* an entity instance of type *Arachidonic_Acid* with the following attributes:

locale = endoplasmic_reticulum, quantity = 1, liberated = true

The test program then retrieves the running instance of *COX1Service*, passes both the above entity instance and the service instance to a service invoker. The service invoker first looks up the capability section of the service to see if the entity instance satisfies the precondition. If so, it retrieves the grounding in the interface section of the WSML service. In this case, it is:

*http://servicemining.org:8001/COX1?wsdl#wsdl.interfaceMessageReference(COX1/produce PGG2/in0)*

The service invoker then gets from the grounding the operation name of the WSDL service, *producePGG2* in this case, and looks up what the corresponding *type* attribute is for *InputSubstance*, the input to the operation. For *producePGG2*, *type* = *ArachidonicAcid*. The service invoker then calls a lowering adapter to package the input SOAP message to the WSDL service operation based on values contained in the entity instance and subsequently invokes the corresponding WSDL service operation. Figure 5.12 shows both the input and output SOAP messages of the *producePGG2* operation. Note that in addition to the translation from the ontological type of the entity instance being *Arachidonic_Acid* to the *type* attribute in the input SOAP message being *ArachidonicAcid*, the following translations have also taken place in the lowering adapter: *locale* to *location*, *quantity* to *amount*, and *liberated* to *flag*. For simplicity, the lowering adapter expects only one extra attribute of *_boolean* type and converts it to a generic boolean *flag*.

```
<?xml version="1.0" encoding="UTF-8"?>
<SOAP-ENV:Envelope xmlns:SOAP-ENV="http://
schemas.xmlsoap.org/soap/envelope/">
    <SOAP-ENV:Header/>
    <SOAP-ENV:Body>
        <q0:producePGG2 xmlns:q0="http://servicemining.org/">
            <q0:ArachidonicAcid>
                <type>ArachidonicAcid</type>
                <location>endoplasmic_reticulum</location>
                <amount>1.0</amount>
                <flag>true</flag>
            </q0:ArachidonicAcid>
        </q0:producePGG2>
    </SOAP-ENV:Body>
</SOAP-ENV:Envelope>
```

(a) Input

```
<?xml version="1.0" encoding="UTF-8"?>
<soapenv:Envelope xmlns:ns1="http://servicemining.org/"
    xmlns:soapenv="http://schemas.xmlsoap.org/soap/envelope/"
xmlns:xsd="http://www.w3.org/2001/XMLSchema">
    <soapenv:Body>
        <ns1:producePGG2Response>
            <ns1:ProstaglandinG2>
                <type>ProstaglandinG2</type>
                <location>platelet</location>
                <amount>1.0</amount>
                <flag>false</flag>
            </ns1:ProstaglandinG2>
        </ns1:producePGG2Response>
    </soapenv:Body>
</soapenv:Envelope>
```

(b) Output

**Fig. 5.12** SOAP Messages of COX1 WSDL Service Operation *producePGG2*

Based on the output SOAP message after the invocation of the WSDL service operation, the service invoker relies on a lifting adapter, which looks up the adapter ontology shown in Figure 5.7, to generate an entity instance of the corresponding ontological type, *PGG2* in this case from the bottom of Figure 5.7. In an end-to-end simulation, such an entity instance can be used as either an input to another operation (e.g., *producePGH2* of *PeroxidaseService*) or an service providing entity if the ontological type corresponds to those that provide service(s).

### 5.4.0.5  Scope Specification

The scope specification (Section 4.1.1) of our service miner is a simple step involving grouping a subset of the ontologies as the mining context. In our experiment, we included in the mining context all seven ontologies as shown in Figure 5.11. In addition, no locale is explicitly specified. Thus, all locales are considered.

### 5.4.0.6  Focused Library Generation

As described in Section 4.1.2, a focused library can be generated based on the mining context. For the experiment, we rely on service interrogation APIs of the WSMX runtime library as indicated in Figure 5.2 to find WSML services that refer to each of the seven ontologies. These services are collected as part of the focused library for later processing.

### 5.4.0.7  Screening

Using the focused library as input, the screening phase first uses the filtering algorithms that are presented in Section 4.2.1 to generate a collection of service composition leads. Figure 5.13 (a) shows examples of service composition leads that are rendered in a Java renderer that we have developed. These leads are then verified using the static verification algorithms presented in Section 4.2.2. Figure 5.13 (b) shows examples where invalid service composition leads have been filtered out by static verification. We show four examples of false composition that are eliminated after static verification.

1. Indirect recognition between *BrainService:processPain* and *Nociceptor_2_Service:transmitPain*. This is invalid because the postcondition of *transmitPain* (see Table 5.3) conflicts with the precondition of *processPain* (see Table 5.3).
2. Indirect recognition between *Nociceptor_3_Service:senseRelief* and *BrainService:processPain*. This is invalid because the postcondition of *processPain* conflicts with the precondition of *senseRelief*.
3. Inhibition from *NF_kappaB_Rel_1_Service:translocate* to *NF_kappaB_Rel_2_Service*. This is invalid because the precondition of

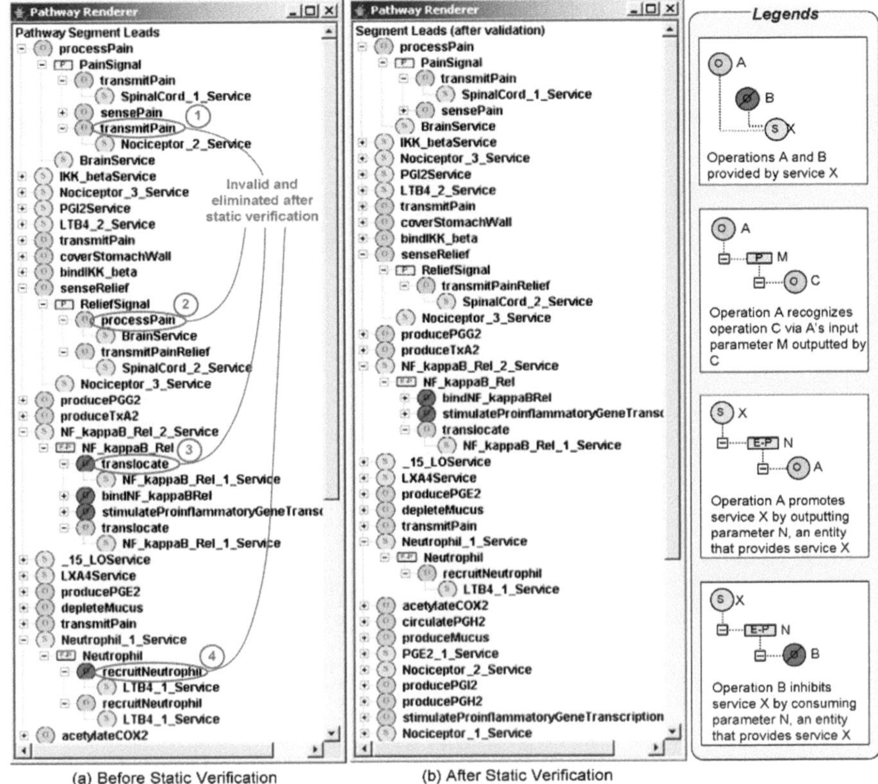







(a) Before Static Verification    (b) After Static Verification

**Fig. 5.13** Example Composition Leads before and after Static Verification

*translocate* conflicts with that of *NF_kappaB_Rel_2_Service*'s operation *stimulateProinflammatoryGeneTranscription*.

4. Inhibition from *LTB4_1_Service:recruitNeutrophil* to *Neutrophil_1_Service*. This is invalid because the postcondition of *recuritNeutrophil* overlaps with the only precondition of *Neutrophil_1_Service*'s operation *produceLTB4* (see line 22 of Algorithm 10).

The linking algorithms (Section 4.2.3) are then applied to the verified leads to generate more comprehensive composition leads as shown in Figure 5.14.

### 5.4.0.8 Evaluation

Objective evaluation is used to identify interesting segments of a lead composition network (see Section 4.3.2.1). In [108], we represented pathways discovered in the screening phase using the tree format (the same as in Figure 5.14) due to its

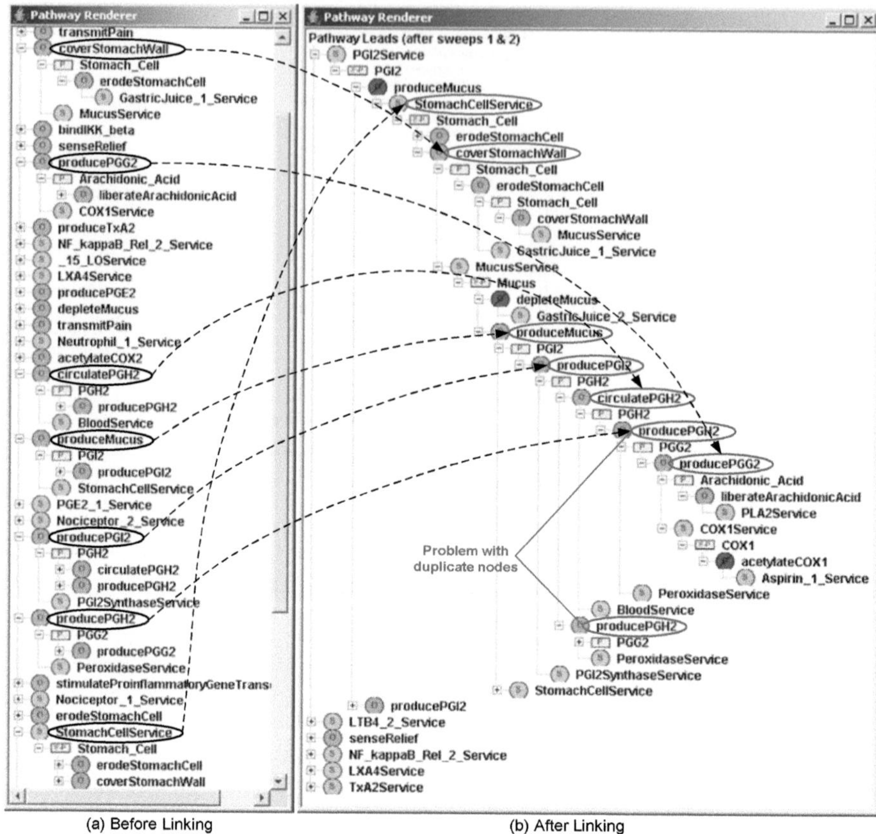

(a) Before Linking      (b) After Linking

**Fig. 5.14**  Example Composition Leads before and after Linking

simplicity in implementation. However, this representation strategy has the inherent difficulty of merging potentially duplicate nodes in these pathways. Figure 5.14 gives an example of two *producePGH2* duplicating each other. We have since extended our rendering algorithms to represent pathways in Graph Markup Language (GraphML) [4], which can then be rendered and automatically arranged using yEd [17]. In Figure 5.15, we show pathways discovered using our adapted screening algorithms and then represented in the graph format. Note that interesting segments in the pathway network are automatically identified and highlighted based on interestingness measure discussed in Section 4.3.1.5. For simplicity, we use only the *novelty* element of interestingness when determining whether a segment is interesting. The novelty is based on the outcome of comparing the source indicator of linkages in the pathway graph representing three types of service/operation recognitions: *promotion*, *inhibition* and *indirect recognition*, as described in Section 3.3. These interesting segments highlight previously hidden linkages between individual

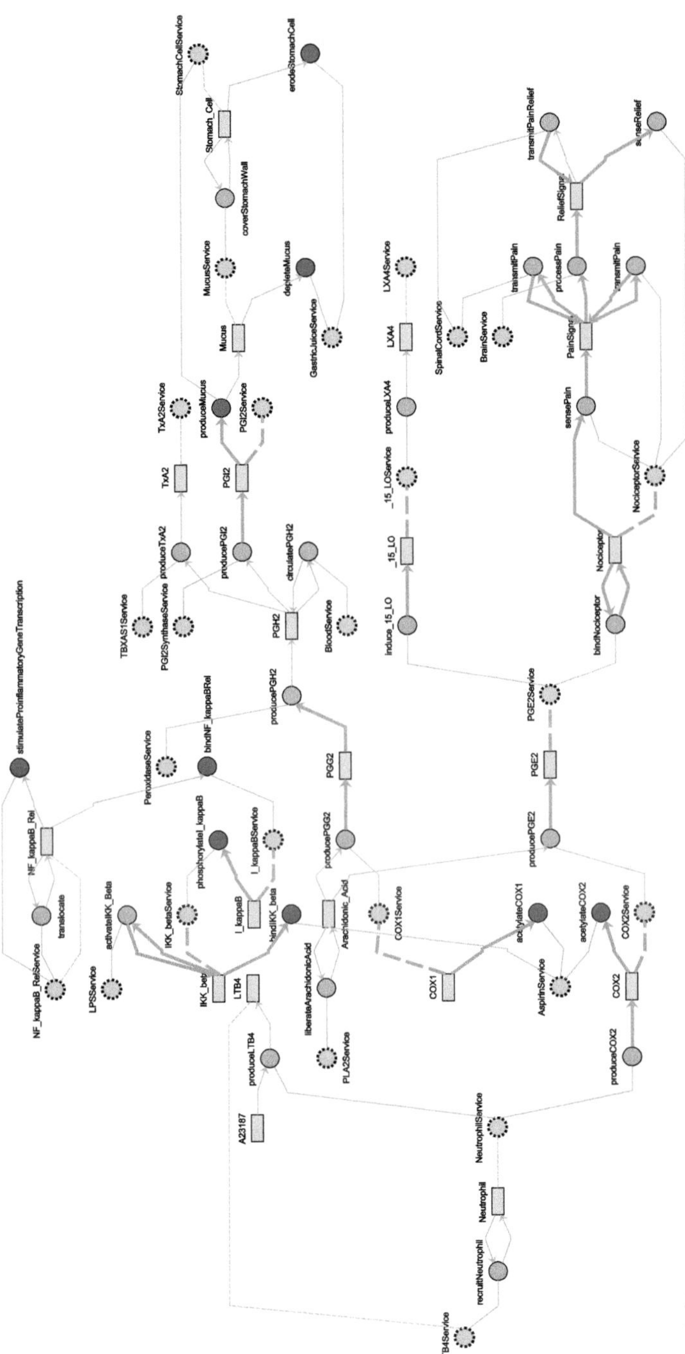

**Fig. 5.15**  Discovered Pathways Represented in GraphML and Rendered in yEd

services and operations. For example, the diagram shows that operation *produce-COX2* of *NeutrophilService* can generate a service providing entity COX2, which may be blocked by operation *acetylateCOX2* of *AspirinService*. Such information is not obviously apparent if we examine simple pathways (see Figure 5.1) that are manually put together individually and independently. For brevity, we display only shortened names for nodes in the graph. We keep the full name containing either the ontological path for entity nodes or the WSML service path for both service and operation nodes in a separate description field (not shown in Figure 5.15). In addition, we omit pre- and post-condition details of operation linking edges such as the two forming a loop between operation *coverStomachWall* and entity *Stomach_Cell* [1]. However, we keep track of the pre- and post-conditions in our service mining client as such information along with the ontological entity paths and WSML service paths are needed when we try to invoke these services during simulation. To ensure the correctness of our algorithms, we compared segments within the automatically discovered pathway network with those constructed manually in Figure 5.1 and found them to be consistent in all cases.

Figure 5.16 shows that based on highlighted interesting segments, the user has selected five nodes of interest, namely, service node *AspirinService*, parameter node *Mucus*, service providing entity node *Stomach_Cell*, parameter node *PainSignal*, and parameter node *ReliefSignal*. The graph expansion algorithms presented in Section 4.3.2.2 are then used to establish a fully connected graph, as shown in Figure 5.17, connecting user-identified interesting nodes with as many interesting segments as possible within a lead composition network. When presented with this highlighted pathway, the user may now hypothesize that an increase in the dosage amount of Aspirin will lead to the relief of pain (usefulness), but may increase the risk of ulcer in the stomach (side effect). Such hypothesis can be tested out using the next simulation step.

### 5.4.0.9  Simulation of Pathways

Although a fully connected subgraph of the pathway network is the basis for the user hypothesis, the simulation itself has to be carried out involving all the constituents (i.e., entities, services and operations) of the network. We outline our simulation strategies in Algorithm 17.

When an operation is to be invoked, the algorithm checks two factors. First, it examines whether all the pre-conditions of the operation are met. An operation that does not have available input entities meeting its preconditions should simply not be invoked. Second, the algorithm determines how many instances are available for providing the corresponding service. This factor is needed because each biological entity of the same type has a discrete service process that deals with input and out-

---

[1] The precondition along the upper edge states that Stomach_Cell is not covered by Mucus and the postcondition along the lower edge states that Stomach_Cell is covered by Mucus.

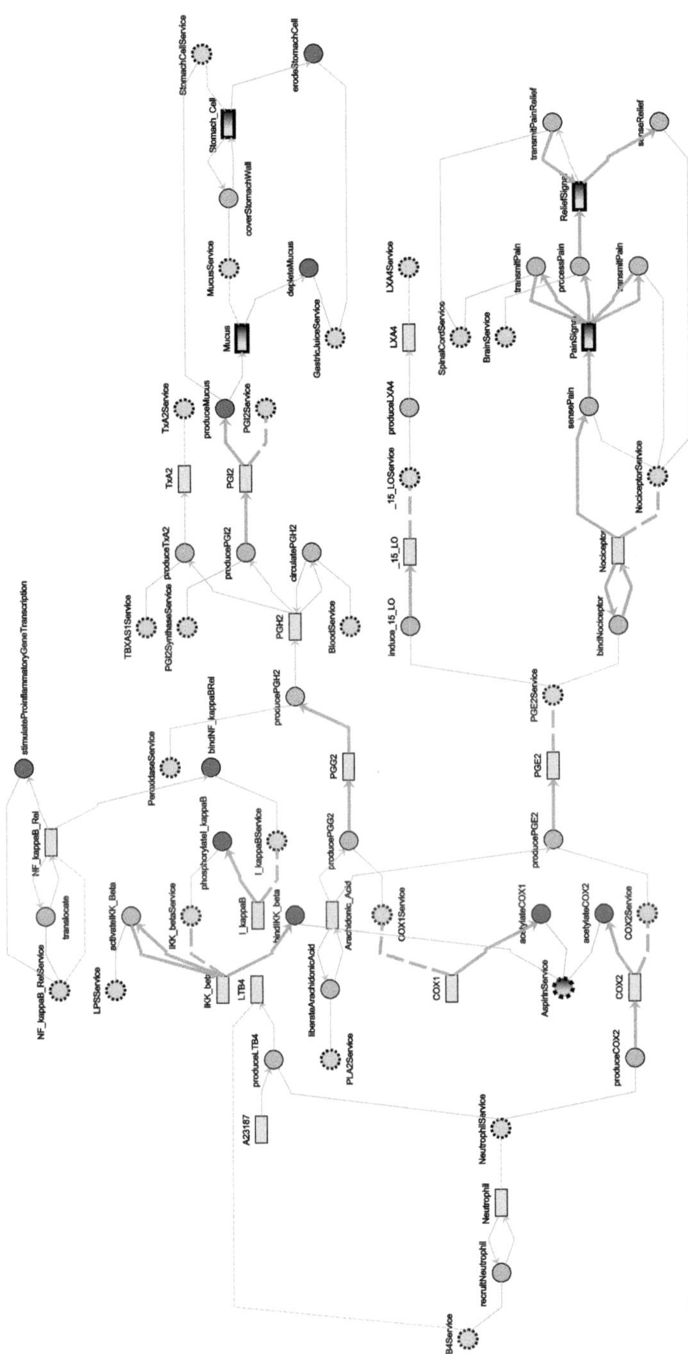

**Fig. 5.16**  User Picks Interesting Nodes in Discovered Pathways

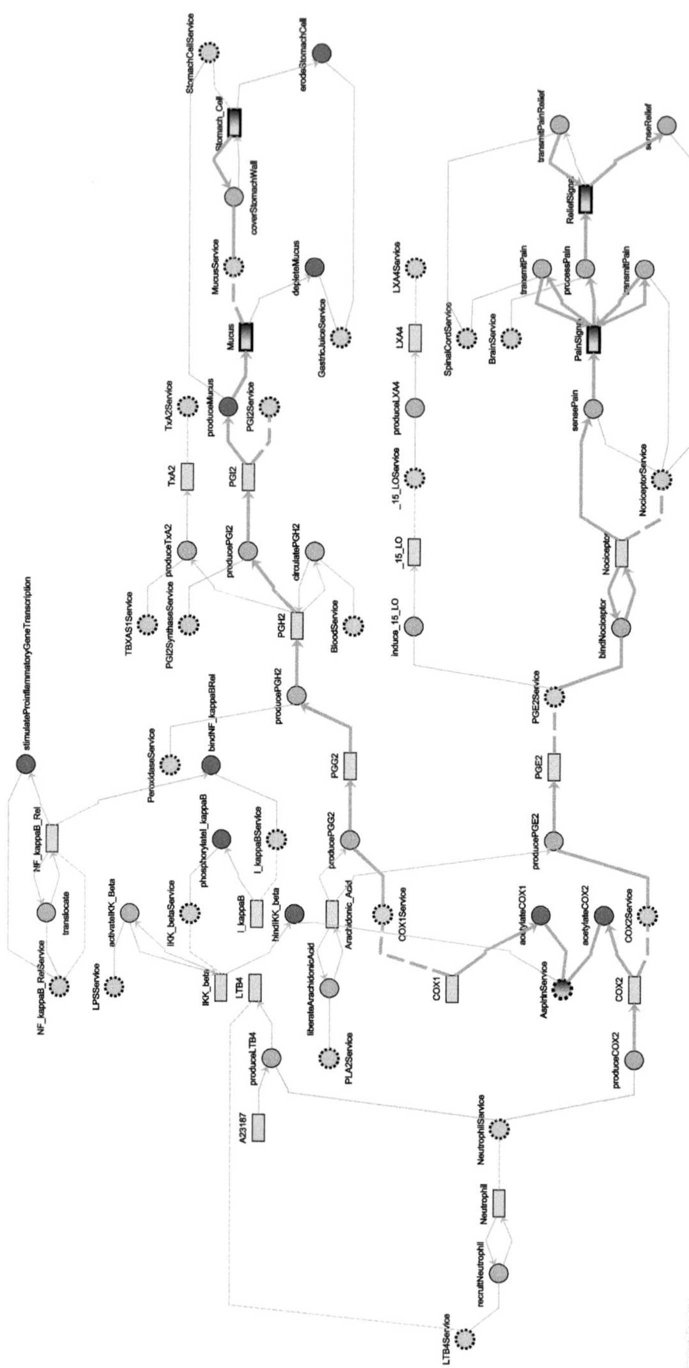

**Fig. 5.17** Interesting Nodes and Edges are Connected

---

**Algorithm 17** Simulation Algorithm

---

**Input**: Pathway Network *PN*, function $f()$ determining initial number of instances for an entity type, total number of iterations $I$, upper bound $S$ for random number generator *random* with uniform distribution
**Output**: Statistics *Stats*
**Variables**: entity type *et*, entity instance container *Container(et)* of type *et*, operation *op*, input entity $op_{in}$, output entity $op_{out}$ and precondition $op_{pre}$ of *op*

1:  **for all** $et \in PN$ **do**
2:    $Container(et) \leftarrow$ create $f(et)$ instances;
3:  **end for**
4:  $Stats \leftarrow$ Tally entity quantities in each container;
5:  **for** $i = 0$ to $I$ **do**
6:    **for all** $op \in PN$ **do**
7:      $s \leftarrow op.getProviderServce();$
8:      $et_{parameter} \leftarrow op.getInputParameter().getEntityType();$
9:      $et_{provider} \leftarrow s.getProviderEntityType();$
10:     **if** $et_{parameter} = et_{provider}$ **then**
11:       $n \leftarrow$ number of entities of type $et_{provider}$ that match $op_{pre}$
12:     **else**
13:       $n \leftarrow$ number of entities of type $et_{parameter}$
14:     **end if**
15:     $n \leftarrow n/S + ((random.nextInt(S) < (n \ modulo \ S))?1 : 0);$
16:     **for** $j = 0$ to $n$ **do**
17:       **if** $\exists op_{in} \in Container(et_{parameter}) : op_{in}$ *matches* $op_{pre}$ **then**
18:         $op_{out} \leftarrow invoke(op)$ *with* $op_{in};$
19:         **if** $et_{parameter} \neq et_{provider} \wedge$ provider is consumable **then**
20:           $Container(et_{provider}).remove(0);$
21:         **end if**
22:         **if** $et_{parameter} \neq et_{provider} \vee$ provider is consumable **then**
23:           $Container(et_{parameter}).remove(op_{in});$
24:         **end if**
25:         $et_{parameter} \leftarrow op_{out}.getEntityType();$
26:         $Container(et_{parameter}).add(op_{out});$
27:       **end if**
28:     **end for**
29:    **end for**
30:    $Stats \leftarrow$ Tally entity quantities in each container;
31: **end for**

---

put of a finite proportion [2]. The available instances of a particular service providing entity will drive the amount of various other entities they may consume and/or produce. For this reason, the algorithm treats each entity node in a pathway network such as the one shown in Figure 5.17 as a container of entity instances of the noted ontology type. In some cases, the service provider is also used as an input parameter. For example, the *sensePain* operation from the *NociceptorService* in Figure 5.1 (g) has a precondition stating that the *Nociceptor* itself should be bound in order to provide this service. In order to express this precondition, we decided to include the

---

[2] This differs from traditional business service processes that are often represented as collective singletons for a given organization (e.g., credit check, loan approval).

service providing entity also as an input parameter. In such a case, the number of service providing instances will be determined by checking further whether each of the service providing entity instances also meets the precondition of the corresponding operation.

In Algorithm 17, an initial number of instances for each entity type $et$ are first generated based on function $f(et)$ (lines 01-03). It is conceivable that an expert may want to create different number of instances at the beginning for different entity types. Next, we conduct $I$ iterations of operation invocations (lines 05-31). We take a snapshot of the quantities at the end of each iteration and before the very first iteration (lines 30 and 04). We determine the number of times the corresponding operation should be invoked based on the quantity of the corresponding service providing entity (lines 7 to 15). To make sure that an operation from a service providing entity of a small quantity also gets the chance to be invoked, a random number generator is used (line 15). Upon invocation of the operation, we remove corresponding entity instance based on the truth table depicted in Figure 5.18. When we determine the provider should be removed (lines 19 to 21), we remove the first instance found in the corresponding container. Since the provider is not the input parameter, it is consequently not involved in the evaluation of the operation precondition. Thus we can remove any one instance found in the container. Lines 22 to 24 are for removing the input parameter instance when the corresponding condition is met. Finally, we add the output parameter instance to the corresponding entity container (lines 25 and 26). In Section 5.4.0.9, we present results from our simulation.

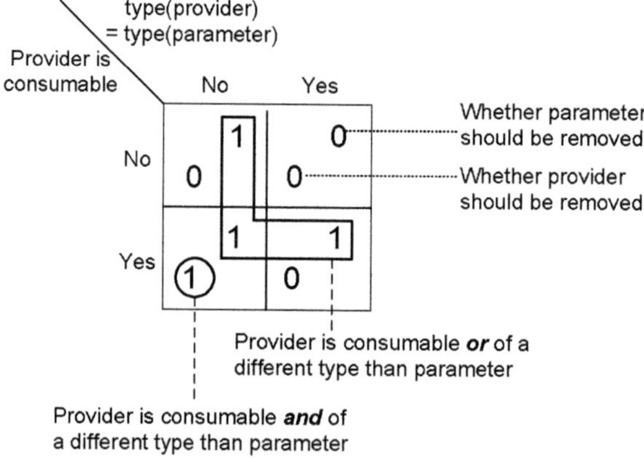


**Fig. 5.18**  Logic for Removing Entity Instance After Operation Invocation

We list in Table 5.4 values for $f(et)$ used in Algorithm 17. In practice, we expect the mining engineer to study the highlighted pathway and use that to determine what the combination of these initial quantities should be before starting the simulation. Note that in cases where there is a singleton entity such as *Brain* and *Spinalcord*, the

**Table 5.4** Experiment Settings

| Entity Type *et* | Initial Quantity $f(et)$ | Providing Service | Provider Consumable |
|---|---|---|---|
| Arachidonic_Acid | 300 | | |
| Aspirin | 10, 40, 70, 100 (Figure 5.19 (a), (b), (c) and (d)) | *acetylateCOX1* | false |
| | | *acetylateCOX2* | false |
| | | *bindIKK_beta* | false |
| Blood | 90 | *circulatePGH2* | false |
| Brain | 30 (Figure 5.19), 15 (Figure 5.20) | *processPain* | false |
| COX1 | 30, 60, 90, 120 (Figure 5.19 (e), (f), (g) and (h)) | *producePGG2* | false |
| COX2 | 15 | *producePGE2* | false |
| GastricJuice | 30 | *erodeStomachCell* | false |
| | | *depleteMucus* | false |
| I_kappaB | 30 | *bindNF_kappaBRel* | true |
| IKK_beta | 30 | *phosphorylateI_kappaB* | true |
| LPS | 30 | *activateIKK_Beta* | false |
| LTB4 | 30 | *recruitNeutrophil* | false |
| | | *incite_inflammation* | false |
| LXA4 | 30 | *suppressInflammation* | false |
| Mucus | 30 | *coverStomachWall* | true |
| Neutrophil | 30 | *produceLTB4* | false |
| | | *produceCOX2* | false |
| NF_kappaB_Rel | 30 | *translocate* | true |
| | | *stimulateProinflammatory-GeneTranscription* | false |
| Nociceptor | 30 | *sensePain* | false |
| | | *transmitPain* | false |
| | | *senseRelief* (disabled for Figure 5.20 (e), (f), (g) and (h)) | false |
| PainSignal | 0 | | |
| Peroxidase | 60 | *producePGH2* | false |
| PGE2 | 0 | *induce_15_LO* | false |
| | | *bindNociceptor* | true |
| PGI2 | 30 | *suppressPlateletAggregation* | false |
| PGI2Synthase | 60 | *producePGI2* | false |
| PLA2 | 30 | *liberateArachidonicAcid* | false |
| ReliefSignal | 0 | | |
| SpinalCord | 30 | *transmitPain* | false |
| | | *transmitPainRelief* | false |
| Stomach_Cell | 60 | *produceMucus* | false |
| TBXAS1 | 15 | *produceTxA2* | false |
| TxA2 | 30 | *vasoconstriction* | false |

initial quantity is only used to help determine how often the corresponding service operation should be invoked. It by no means indicates that there is more than one of such entity. We start by simulating how the quantity of Aspirin affects the erosion of stomach and sensation of pain. The simulation results obtained from each run of the mining client are compiled into an Excel spreadsheet, which is then used to generate plots such as those in Figures 5.19 and 5.20.

Figure 5.19 (a), (b), (c) and (d) show that given a fixed initial quantity of 60 for COX1, the increase in the dosage amount of Aspirin has a negative effect on the stomach, i.e., as the quantity of Aspirin continues to increase from (a) to (d), the severity of stomach erosion also increases. For example, when the quantity of Aspirin is 10, there is no sign of stomach erosion. When the quantity of Aspirin increases to 100, the quantity of stomach cell, after 150 iterations of operation invocation, drops to 20, which is one third of the initial quantity. This confirms the user hypothesis that Aspirin has a side effect on the stomach. In addition, we also noticed that given a fixed quantity of Aspirin, the reduction of the initial quantity of *COX1* also has a negative effect on the stomach (Figure 5.19 (e), (f), (g) and (h)). When the initial quantity of *COX1* is high, it takes longer for all the *COX1* to get acetylated by Aspirin. As a result, enough *PGG2* and consequently *PGH2* and *PGI2* will be built up to feed into the *produceMucus* operation of the *StomachCellService*. As the initial quantity of *COX1* becomes smaller and while the depletion rate of *Mucus* by *GastricJuiceService* remains the same, less *Mucus* is being produced by the *StomachCellService* as less *PGI2* becomes available.

While Figure 5.19 clearly illustrates the relationships between Aspirin and *Stomach_Cell*, the relationship between the dosage amount of Aspirin and the sensation of pain is less obvious in the same Figure. Except for Figure 5.19 (a), which shows some accumulation of *PainSignal* when the quantity of Aspirin is 10, the rest of plots show no pattern of such accumulation or the variation thereof. A closer look at the highlighted pathway in Figure 5.17 reveals that this is actually consistent with the way the simulation is set up. Since *PainSignal* is created and then converted by the *Brain* to *ReliefSignal*, which disappears after it is sensed by *Nociceptor*, this whole path at the bottom actually acts as a 'leaky bucket'. To examine exactly what is going on along that path, we decided to make two changes in the simulation setting. First, we reduced the quantity assigned to *Brain* from 30 to 15. This creates a potential imbalance between the production rate of *PainSignal* and *ReliefSignal* since the *processPain* operation from the *BrainService* will be consequently invoked less frequently than the *sensePain* operation from the *NociceptorService*. Second, we disabled the *senseRelief* operation of the *NociceptorService*. This essentially stops the leaking of the *ReliefSignal* that are generated as a result of the *PainSignal*. When we apply only the first change to the simulation, the imbalance of the processing rates for *PainSignal* and *ReliefSignal* results in a net accumulation of *PainSignal* when the quantity of Aspirin is 10 (Figure 5.20 (a)). When the quantity is increased to 40 (Figure 5.20 (b)), we see there are some occasional and temporary accumulation of *PainSignal*. As the quantity of Aspirin continues to increase (Figure 5.20 (c) and (d)), we see no detectable accumulation of *PainSignal*. Finally, we apply the second change along with the first one. Consequently, we notice that while the pattern of

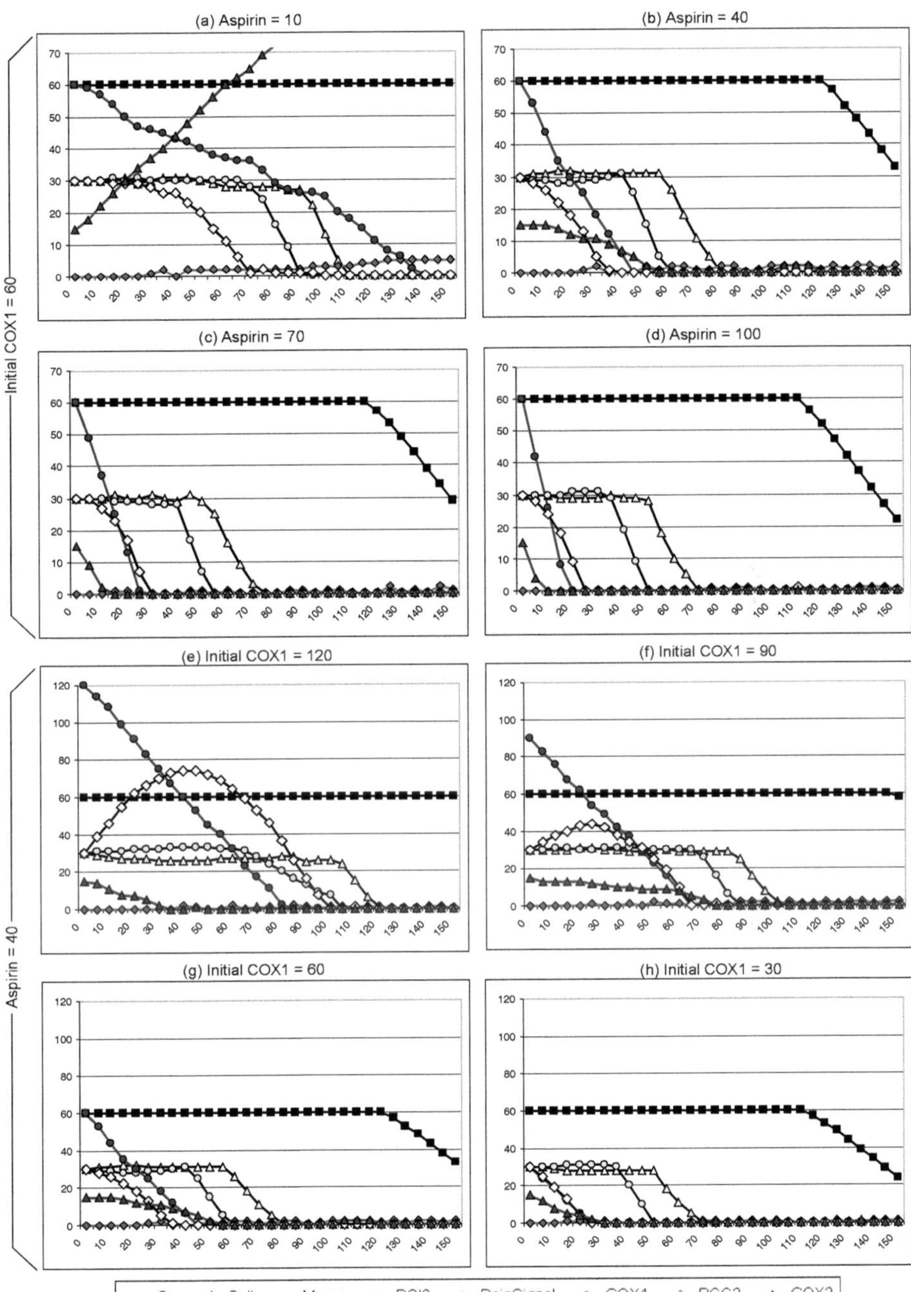

Fig. 5.19  Simulation Results with Original Configurations

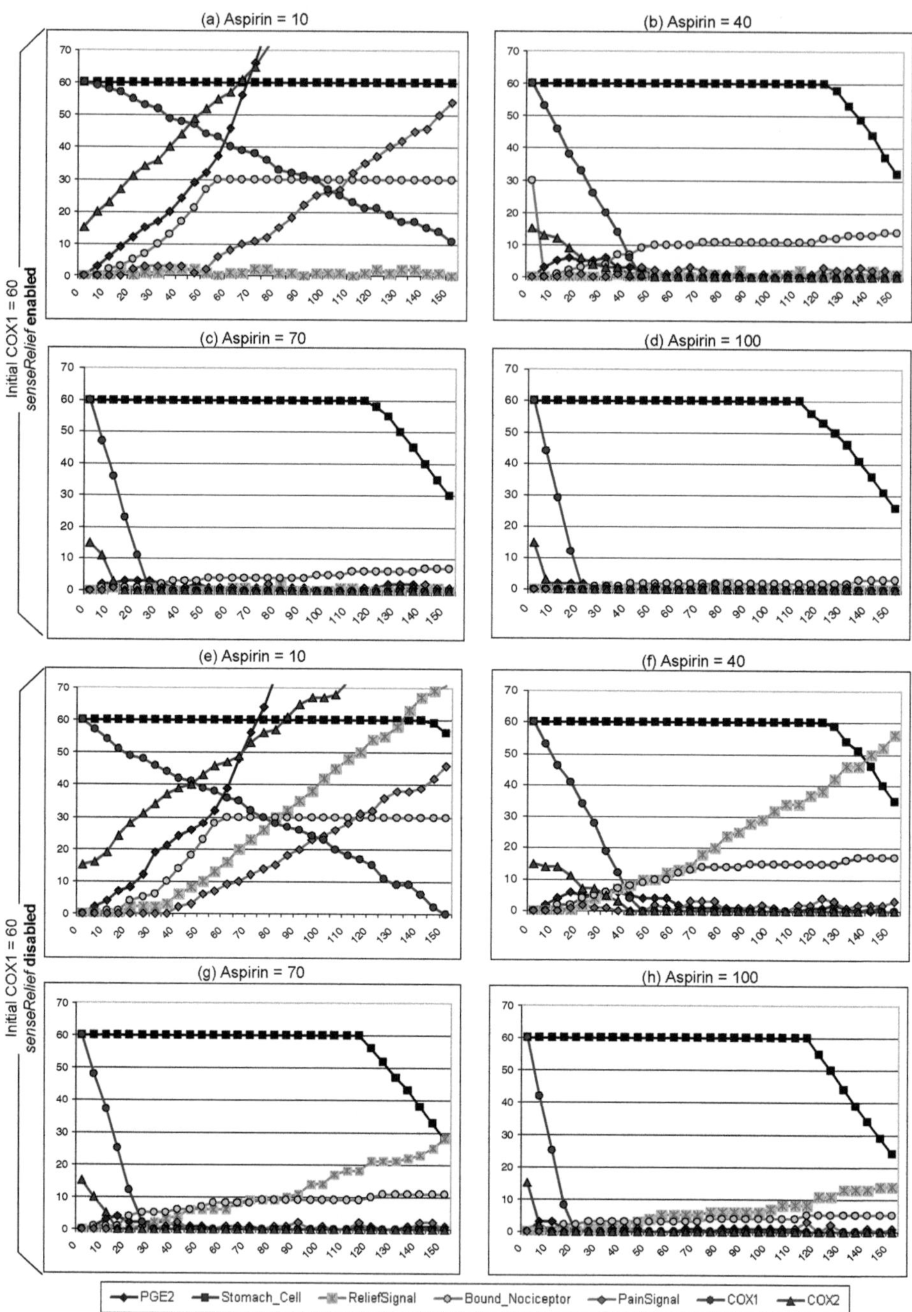


**Fig. 5.20** Simulation Results with Modified Configurations

*PainSignal*'s accumulation hasn't changed much, there is a consistent accumulation of *ReliefSignal*. Since each *PainSignal* is eventually converted to a *ReliefSignal* by the *Brain* according to the highlighted pathway in Figure 5.17, the rate of *ReliefSignal*'s accumulation actually provides a much better picture on how fast *PainSignal* has been generated. We see that as the dosage amount of Aspirin increases, less *ReliefSignal* is generated, an indication that less *PainSignal* has been generated. Thus it is obvious that the increase of the dosage amount of Aspirin has a positive effect on the suppression of *PainSignal*'s generation. This confirms the other half of user's original hypothesis.

Simulation results such as these presented in Figures 5.19 and 5.20 provide information to a pathway analyst who would otherwise get such information from *in vitro* and/or *in vivo* exploratory mechanisms. Using the service-oriented simulation environment, the interrelationships among various entities involved in the pathway network can now be exposed in a more holistic fashion than traditional text-based pathway discovery mechanisms, which inherently lack the simulation capability.

Kaufmann's accumulation hasn't changed much, there is a considerable amount of [illegible] of learning spent. Since each [illegible] is eventually connected to a Reply/Input in the Right according to the highlighted paths by (by Figure 3.7), a value of Reply [illegible] accumulation already provides a much better picture of how each Reply has been generated. We see that as the [illegible] amount of every increases, less AP (Reply/[illegible]) is presented, an indication that less AP/Signal has been generating. This [illegible] obtains that the nature of the [illegible] amount of Reply/[illegible] has a possible effect on the expression of [illegible] generation. This confirms the expectation that used original Reply/[illegible].

Simulation results and [illegible] thus presented in [illegible]. [illegible] [illegible] exploratory simulations. Using the results, a channel simulation [illegible] [illegible] [illegible] [illegible] [illegible] in the layer [illegible].

# Chapter 6
# Related Work

In this Chapter, we survey related work. These include Web service composition techniques and standards, data mining, log-based interaction and process mining, and natural computing. We also discuss similarities among and differences between these approaches and the Web service mining approach as presented in this book.

## 6.1 Web Service Composition

Several standardization and prototype efforts were undertaken in the field of Web service specification and composition. To illustrate the relationships between many of the approaches that resulted from these efforts, we refer to Figure 6.1 ([88, 11, 14]), which organizes various Web service related standards, including those related to Web service composition, according to their levels of abstraction.

Composition related approaches (Figure 6.1) can be grouped into two different categories [88]: *business process-oriented* and *semantics-based*. Several important approaches (e.g. Petri-nets and algebraic process composition, WSFL, XLANG) are not shown in Figure 6.1. They either serve as the formal basis for other approaches or are merged into other standards (e.g., BPEL4WS). We overview each of these in the following sections.

### 6.1.1 Business Process-Oriented Web Service Composition

Business process-oriented approaches include: Petri-nets, algebraic process composition, *WSFL, WSCI, XLANG, BPEL4WS*, Business Transaction Protocol (BTP), Business Process Modeling Language (BPML) and Electronic Business XML (ebXML).

G. Zheng and A. Bouguettaya, *Web Service Mining: Application to Discoveries of Biological Pathways*, DOI 10.1007/978-1-4419-6539-4_6,
© Springer Science+Business Media, LLC 2010

UDDI | ebXML Registries | *Discovery*

| ebXML CPA | *Contracts and Agreements*

WSMX | OWL-S Service Model | BPEL4WS | BPML | *Process and workflow Orchestrations*

WS-AtomicTransaction and WS-BussinessActivity | BTP | *QoS: Transactions*

WSML Full | OWL-S Service Profile | WS-Reliable Messaging | WS-Coordination | WSCI | ebXML BPSS | *QoS: Choreography*

WSML Rule | WSML DL | OWL-S Service Grounding | WS-Security | WSCL | *QoS: Conversations*

WSML Core | OWL | PSL | WS-Policy | WSDL | ebXML CPP | *QoS: Service Descriptions and Bindings*

RDF | SOAP | ebXML Messaging | *Messaging*

XML, DTD, and XML Schema | *Encoding*

HTTP, FTP, SMTP, SIP, etc. | *Transport*

☐ Semantics Oriented      ☐ Business Process Oriented

**Fig. 6.1** Relationships among Proposed Standards and Methodologies

### 6.1.1.1 Petri-nets

The Petri-net approach [48] graphically represents a process as a directed and connected graph where some nodes (referred to as places) can be used to represent states and other nodes (referred to as transitions) can be used to represent operations. When there is at least one token in every place connected to an operation, the operation is enabled. An enabled operation might fire by removing one token from every input place, and depositing one token in each output place. At any given time, a service can be in one of the following states: *not instantiated, ready, running, suspended,* or *completed.* After a net for each service is defined, composition operators can perform a variety of compositions including *sequence, alternative, iteration, parallel,* etc. A process can be analyzed in a number of ways using Petri-nets due to its abundance of analysis techniques [33, 37, 66, 96]. For example, Petri-nets can be used to determine the absence of deadlocks or livelocks (whether composition will terminate in a finite number of steps).

### 6.1.1.2 Algebraic Process Composition

Algebraic service composition models services as mobile processes. Mobile processes theory is based on $\pi$-calculus [75], in which the basic entity can be one of the following processes: *empty process*, a choice between several I/O operations, a parallel composition, a recursive definition, or a recursive invocation. I/O operations can be input (receive) or output (send). For example, $x(y)$ denotes receiving tuple $y$ on channel $x$; $\bar{x}[y]$ denotes sending tuple $y$ on channel $x$. Dotted notation specifies an action sequence, such as $\bar{u}[1,v].v(x,y,z).\bar{u}[x+y+z]$, in which a process sends tuple $[1,v]$ on channel $u$, then receives a tuple at channel $v$ whose components are bound to variables $x$, $y$, and $z$, and finally sends the sum of $x+y+z$ to channel $u$. Parallel process composition is denoted with $A|B$. Describing services in such an abstract way allows for the reasoning about the composition's correctness [72].

### 6.1.1.3 WSFL

IBM's Web Services Flow Language (WSFL) [15, 65] is an XML language for describing Web service compositions. WSFL is based on Petri-nets [88]. It provides two types of model, *flow* model and *global* model. The flow model aims to specify the logic of a business process. It uses a directed graph to model the execution sequence of the functionality provided by a composed service as a flow of control and data between component services. Each node of the graph, referred to as an *activity*, represents a step of the business goal the composition tries to achieve. The types of links are used to connect activities. The first type, referred to as *control* links, prescribe the order for performing the activities. The second type, referred to as *data* links, represent the flow of data between activities. The global model aims to define the mutual exploitation of Web services of decentralized or distributed participants in a business process. No specification of an execution sequence is provided. Instead, it relies on the use of *plug* links to represent the interactions among participants. WSFL also aims to support the recursive composition of services. In WSFL, every Web service composition (a flow model as well as a global model) can itself become a new Web Service, and can thus be used as a component of new compositions.

### 6.1.1.4 WSCI

BEA Systems' Web Service Choreography Interface (WSCI) [97] is an XML-based interface description language used to describe how Web Service operations, such as those defined by WSDL, can be choreographed in the context of a message exchange in which the Web Service participates. WSCI describes how the choreography of these operations should expose relevant information, such as message correlation, exception handling, and transaction description. Message correlation is realized by associating exchanged messages with correlation properties identifying a conversion. Exception can be due to the receipt of a out-of-context message, occurrence of a fault or a timeout. A transaction is used to group a set of activities that should be executed in an all-or-nothing manner. WSCI describes the behavior of Web service in terms of choreographed activities. Activities may be atomic or complex, i.e. recursively composed of other activities. Choreography describes temporal and/or logical dependencies among activities. Types of choreographies supported by WSCI include: *sequential execution, parallel execution, looping,* and *conditional execution.*

### 6.1.1.5  XLANG

Microsoft's XLANG [92] is based on $\pi$-calculus [88]. It extends a WSDL service description with a behavioral element at the intra-service level. A behavior defines the list of actions that belong to the service and the order in which these actions must be performed. XLANG supports two types of actions: WSDL operations and XLANG-specific actions including exceptions and deadline/duration based timeouts. Similar to WSCI, message correlation and transaction are also supported in XLANG. At the inter-service level, XLANG defines a *contract* detailing the connections between service ports used to stitch together individual service descriptions.

### 6.1.1.6  BPEL4WS

Due to the incompatibility between WSFL and WSCI/XLANG, researchers developed Business Process Execution Language for Web Services (BPEL4WS) [26, 42]. BPEL4WS combines features in WSFL (process graph) and WSCI/XLANG (process structural constructs) to support process-oriented service compositions. A process is composed of *activities* (taken from WSFL and WSCI). The execution of a process may encounter *exceptions* (taken from WSCI and XLANG). Both message correlation and transactions from WSCI and XLANG are also supported in BPEL4WS. There are several BPEL2WS orchestration server implementations for both J2EE and .NET platforms, including IBM WebSphere [54], Oracle BPEL Process Manager [82], Microsoft BizTalk [74], OpenStorm Service Orchestrator [81] and ActiveBPEL [21].

### 6.1.1.7  BTP and BPML

Business Transaction Protocol (BTP) [79] is designed to support interactions that cross application and administrative boundaries. BTP relaxes the ACID properties via two subprotocols: *atoms*, where isolation is relaxed, and *cohesions*, where both isolation and atomicity are relaxed. Business Process Modeling Language (BPML) [78] shares the same root in WSCI with BPEL4WS. It uses WSCI for expressing public interfaces and choreographies. BPML also provides advanced process model semantics such as nested processes and complex compensated transactions.

### 6.1.1.8 ebXML

Electronic Business XML (ebXML) [95] is an international initiative established by the United Nations Centre for Trade Facilitation and Electronic Business (UN/CEFACT) and the Organization for the Advancement of Structured Information Standards (OASIS). ebXML aims at defining standard business processes and trading agreements among different organizations. The vocabulary used for an ebXML specification consists of the following:

- a *process specification document* describing the activities of the parties in an ebXML interaction
- a *collaboration protocol profile* describing an organization's profile (e.g., supported business processes, its roles in those processes, the messages exchanged, and the transport mechanism for the messages)
- a *collaborative partner agreement* representing an agreement between partners

The infrastructure includes an ebXML registry that stores important information about businesses along with the products and services they offer. ebXML registries have an advantage over UDDI (short for Universal Description, Discovery, and Integration) registries in that they allow queries on keywords based on Structured Query Language (SQL).

## 6.1.2 Semantics-based Web Service Composition

Semantics-based Web service composition approaches take advantage of service ontologies when describing Web service descriptions. The use of such ontologies allows for unambiguous interpretation of Web service semantics by a computer. Based on the Defense Advanced Research Projects Agency (DARPA) Agent Markup Language Ontology Inference Layer (DAML+OIL), Web Ontology Language for Services (OWL-S) [99], previously known as DAML-S, is a service ontology that aims at enabling Web services reasoning, composition planning and automatic use of services by software agents. OWL-S models services using a three-part ontology: *profile*, *model*, and *grounding*. A service profile is an abstract characterization of what a service can do. This includes the required *input* of the service, the *output* the service will provide to the requester, the *preconditions* and the *effects* of the execution of the service. These are collectively known as the IOPEs of the given service. The service model describes a service's behavior as subprocesses using a process graph. There are three kinds of subprocesses in a graph: *atomic processes* that have no subprocesses and are invocable, *simple processes* that do not have subprocesses but might be invocable, and *composite processes* that have subprocesses linked by control constructs, such as *sequence, split, split and join, choice, iteration,* and *if-then-else*. Service grounding describes how to access and use a Web service. It includes descriptions for message formatting, transport mechanisms, protocols, and serializations of types. Different from BPML, which is business-oriented, OWL-S provides the necessary vocabulary and properties that are generic to a process model.

A more recent language for Semantic Web service specification is the Web Service Modeling Language (WSML) [10]. WSML is based on the Web Service Modeling Ontology (WSMO) [11]. The core language of WSML uses description logics and logic programming. While not restricting the expressive power of the language for the expert user, WSML takes consideration of users who are not familiar with formal logic by making the distinction between conceptual and logical modeling. Like OWL-S, WSMO declares inputs, outputs, preconditions, and results associated with services. Unlike OWL-S, WSMO does not provide a notation for building up composite processes in terms of control flow and data flow. Instead, it focuses on specification of internal and external choreography, using an approach based on abstract state machines. A major difference between WSMO and OWL-S is WSMO's emphasis on the production of a reference implementation of an execution environment, WSMX, and on the specification of mediators – mapping programs that solve the interoperation problems between Web services.


### 6.1.3 Contrasting Web Service Mining and traditional Web Service Composition Approaches

Service composition can be carried out using two fundamentally different strategies. The first is the top down strategy, which requires a user to provide a goal containing *specific* search criteria to start the composition process, as do traditional Web service composition approaches. Since the goal provided by the user already implies what type of compositions the user anticipates, the evaluation of composition *interestingness* is not a major concern in these approaches. The second is the bottom up strategy, which is driven by the desire to find *any* interesting and potentially useful compositions of existing Web services without using specific search criteria. We refer to the second strategy as *Web service mining* because of its proactive nature in seeking for composed Web services. In the absence of a top-down query, Web service mining techniques need to address how *interestingness* of service compositions can be determined. In [109, 112], we introduced an interestingness measure used to evaluate composite services discovered via our service mining algorithms. To date, no work has been reported elsewhere in the field of Web service mining that relies on no *a priori* knowledge about the composabilities of Web services. The closest work is found in log-based interaction and process mining, which is discussed in Section 6.3.

## 6.2 Contrasting Web Service Mining and Data Mining

Data mining is the extraction of interesting information from data contained in databases. It relies on computer programs to sift through databases for finding regularities or patterns. Data mining has received much attention due to the increased amount of data that exist and are made available to people through cheap storage. Today enterprises collect up to terabytes of data [100]. Many have looked into using data mining techniques for purposes of data classification, anomaly detection and prediction. Data mining techniques search for consistent patterns and/or systematic relationships between variables [57]. They then validate the findings by applying the detected patterns to new sets of data. These techniques have been used to forecast weather [102], predict stock market [27], sports [49] and electric load fluctuation [80], assist diagnosis [57], and detect fraud [49].

Similar to Web service mining, data mining is driven by the motive of extracting previously unknown and useful information from existing data. Data mining techniques rely on *static data* being the subject of study. Unlike static data, Web services encapsulate **behaviors** through a set of operations that are intrinsically *dynamic* in nature. Consequently, the variations in data values that traditional data mining techniques rely on cannot be used to determine whether two services are composable. It is the syntactic operation structures, the semantics of messages sent across these operations and measures such as quality that determine whether two services are composable. As a result, the task of Web service mining is considerably more challenging.

## 6.3 Log-based Interaction and Process Mining

In [40], Dustdar et. al. rely on analyzing Web service execution log data to discover potential process workflow instances involving these services. While this approach may work well for business processes over time as more execution logs are expected to become available in this area, the challenge of identifying interesting workflows in the absence of such logs, especially at the time when component Web services are just introduced, is still real. Our Web service mining framework allows the mining of interesting composite services to be carried out in the absence of execution logs. When applied to the field of pathway discovery, where the expedience of such discovery is key to success, our approach enables the proactive discovery of interesting pathways upon the availability of these services.

## 6.4 Natural Computing

Natural computing is the field of research that "investigates models and computational techniques inspired by nature and, dually, attempts to understand the world around us in terms of information processing" [60]. In our research, we devised service mining techniques based on molecular recognition and applied these techniques back to the discovery of biological pathways. Consequently, our research can be viewed as a subfield of natural computing. Natural computing covers many other theoretical researches, algorithms and software applications. Of particular relevance to our research are efforts to understand the universe from the point of view of information processing and how they change what we mean by computation. Classical examples of nature-inspired models of computation include *cellular automata*, *neural computation*, and *evolutionary computation*. More recent examples include *swarm intelligence* and *artificial immune systems*.

The notion of cellular automata was first conceived by Ulam and von Neumann in the 1940s. A cellular automaton is a dynamical system consisting of a regular grid of cells in discrete space and time. Each cell can be in one of a finite number of states and transition its state according to a given set of rules. These rules determine the cell's future state based on its current states and those of some of its neighbors. As a result, the configuration of the entire grid of cells would update synchronously according to the transition rules. Cellular automata have been applied to the study of communication, computation, growth, reproduction, competition, and evolution.

The field of neural computation originates from the model of artificial neurons first proposed by McCulloch and Pitts [68]. Neural computation attempts to first help unravel the structure of computation in nervous systems of living organisms, i.e., how the brain works, and use that knowledge to help yield significant computational advances, i.e., how to build an intelligent computer. The quest for the second goal eventually gave rise to the study of artificial neural networks or simply neural networks. Similar to the brain, a neural network can be trained and "learn" to distinguish one pattern from another. Various types of neural networks have been developed. They include the *perceptron* [85], back propagation networks (BPN) based on the learning algorithm introduced by Rumelhart [86], Cognitron [46], and Kohonen's feature maps [62]. In a neural network (e.g., in BPN and Cognitron), layers of nodes are linked by weights between layers, with one of the layers being the input layer and another being the output layer. The remaining layers are hidden between the input and output. The learning is achieved through training the network repeatedly by feeding it with training instances so the weights are optimized. As a result of this optimization, the output layer eventually generates significantly different values for different instances in the training set, an indication that the network has finally recognized the provided instances. The assumption is that the trained network can then be used to recognize instances similar to any of the instances contained in the training set. Neural networks have been used for clustering, feature extraction, and anomaly detection [27]. In bioinformatics, neural networks have been used to cluster gene expression patterns [51] and predict HIV drug resistance [39].

Unlike cellular automata and neural computation that emulate features of a single biological organism, evolutionary computation [24] inspired by the Darwinian evolution theory [34] emulates the dynamics of an entire species of organisms. An artificial evolutionary system is a computational system based on the notion of simulated evolution. It features a constant- or variable-size population of individuals, a fitness criterion according to which the individuals of the population are being evaluated, and generically inspired operators that produce the next generation from the current one. Evolutionary computation has been applied mostly to the study of optimization.

Swarm intelligence [41], sometimes referred to as collective intelligence, is defined as the problem-solving behavior that emerges from the social interaction of a collection of mobile biological organisms (e.g., bacteria, ants, bees, fish, birds) or agents. Two examples are the flocking behavior [84] of birds and the foraging behavior of ant colonies. Particle swarm algorithms inspired by the flocking behavior of birds have been applied to unsupervised learning, game learning, scheduling and planning applications, and design applications. Ant algorithms [38] inspired by the foraging behavior of ant colonies have been used to solve combinatorial optimization problems defined over discrete search spaces.

Artificial immune systems [43, 35] are devised according to the natural immune system of biological organisms. The natural immune system is viewed as an information processing system that performs many complex computations such as the distinction between self and nonself and neutralization of nonself pathogens (e.g., viruses, bacteria, fungi, and parasites). Upon invasion of foreign antigen into the body, a few of body's immune cells can detect and bind to the invaders. This detection also triggers the generation of a large population of cells that produce matching antibodies that aid in the destruction of neutralization of the antigen. The immune system retains some of these specific-antibody-producing cells in the immunological memory, so that any subsequent exposure to a similar antigen can lead to a more rapid immune response. Applications inspired by these include computer virus detection, anomaly detection, fault diagnosis, pattern recognition, machine learning and so on.

In parallel to the above examples that focus on nature-inspired computational models and techniques, a dual direction of research in natural computing attempts to understand processes that take place in nature as information processing. For example, one view of studying the entire genomic regulatory system is to think of it as a "genomic computer" [58] where robustness is achieved by means such as rigorous selection. In this selection, non or (poorly)-functional processes are rapidly degraded by various feedback mechanisms. In such a computer, the distinction between hardware and software breaks down, i.e., the genomic DNA provides both the hardware and the digital regulatory code (software). At the far spectrum of views of nature as computation is the idea of viewing the entire universe as some kind of computational device or a quantum computer [67] that computes itself and its own behavior.

# Chapter 7
# Conclusions

The growing popularity of the Web and Web services has provided a great potential for a new research discipline, aka, Web service mining, as we have identified here. An effective service mining tool would allow potentially interesting and useful composite Web services to be discovered based on the availability of component services, rather than solely on human intuition or knowledge about the composability of these component services, as do traditional top down service composition approaches. We proposed a Web service mining framework that aims at the proactive discovery of interesting and useful service compositions. We also applied the framework to the discovery of useful pathways linking service-oriented models of biological processes. We summarize the main contributions of our research below.

**Web Service Mining Framework** - We identified two main challenges faced in the mining of Web services: combinatorial explosion and the determination of service composition interestingness and usefulness. In searching for effective strategies to address these two challenges, we first explored the similarities between Web services and molecules in the natural world. These similarities led us to studying the drug discovery process. Strategies used in this process inspired the conception of the various phases of our mining framework. To address the first challenge, we relied on a Web service ontology to organize constructs of Web services. We proposed to describe shared capabilities of Web services through operation interface, allowing Web services to declare their role as either as provider or consumer of an operation interface. We derived the concept of service and operation recognition from the phenomenon of molecular recognition. We then developed several publish/subscribe mechanisms centered around the concepts of operation interface and service/operation recognition. These enabled us to convert the traditional top down combinatorial search problem into a more efficient process of bottom up service and operation recognition. We incorporated both static verification in the screening phase and dynamic runtime simulation-based verification in the evaluation phase. To address the second challenge, we established a set of measures that can be used to objectively determine the interestingness and usefulness of service compositions.

G. Zheng and A. Bouguettaya, *Web Service Mining: Application to Discoveries of Biological Pathways*, DOI 10.1007/978-1-4419-6539-4_7,

We also strived to push the expensive subjective evaluation step towards the end of the mining process.

**Web Service Modeling of Biological Processes** - Computer-based representation of biological entities and processes is currently dominated by two approaches: free text description and computer models. The first approach, while easy to carry out, inherently lacks the ability to support computer-based simulation. The second approach has the limitation of models being constructed to simulate entities in isolated local environments, essentially making the sharing and interoperation of these models a difficult task. To address these limitations and to bridge the gap between the two approaches, we proposed to model biological processes as Web services. This approach fosters better sharing of biological models by making them both discoverable and directly invocable on the Web. For experiment, we attempted to model biological processes as WSDL services and expose their semantic interfaces using WSML. We then deployed these services to runtime environment including WSMX and Jetty Web server.

**Pathway Discovery through Web Service Mining** - We applied our Web service mining framework to the WSML service models of biological processes and demonstrated its effectiveness in the automatic discovery of pathway networks linking these models. We developed a set of graph related algorithms identifying interesting pathway segments and the establishment of a fully connected subgraph. These are then provided to user as visual aids and serve as basis for hypothesis formulation. We developed simulation strategies and demonstrated the feasibility of simulating these pathways through invoking service oriented biological models directly on the Web. Based on results from such simulation, user can either confirm or rejects his/her original hypotheses and make the ultimate decision as to whether the pathway under study is really useful.


## Directions for Future Research


The Web is transforming how we search for, discover, acquire, process and use information. Fueling into this transformation is the continuous evolution of the way information is being collected, organized and presented on the Web. From a Web of data to a Semantic Web of both data and services, the Web has experienced explosive growth in the past decade with continuous introduction of new sources of information and continuous development of services with new methods to collect, new models to organize, and new tools to present such information on the Web. The services oriented architecture encourages the sharing of the best-of-the-breed functionalities on the Web and promotes collaboration among organizations and individuals. Such collaboration is realized through interactions among relevant services, consequently generating value-added services. Service interactions for the purpose of service composition can be carried out in two complimentary fashion, top down and bottom up, as we have discussed earlier. We have focused in this book on the second type of service composition, aka Web service mining. In the following, we identify several open research problems related to Web service mining.

*Dynamic Mining of Web Services* – Web services used as input to service mining can themselves change continuously due to, for instance, increase in the visibility of underlying models and external conditions described in Figure 3.6. Such change can potentially invalidate previous mining results or provide potential for obtaining more interesting results. In the context of biological pathway discovery, as service models of biological processes become more mature and detailed, one would expect mining results to reflect more closely real world processes and their interactions that are being pursued. Dynamic service mining would be required to update mining results rapidly or in a timely manner to keep them up with ongoing changes in Web services. To achieve this, different phases of the mining framework may need to be updated concurrently and asynchronously.

*Collaborative Mining of Web Services* – We focused in this research on how independent service mining can be carried out. Since service mining can be performed concurrently by multiple organizations, each within similar or related mining contexts, it can be beneficial to these organizations by teaming up and forming alliances to collaborate on their service mining activities. Further research will be needed to address how to efficiently share service mining results, potentially during various mining phases, among partner organizations and coordinate their service invocations (e.g., for the purpose of validation and predictive analysis). To ensure successful collaboration among partner organizations, one may need to address how to handle the way organizations trust one another so that the collaboration is fair and mutually beneficial.

*Service-based Mining of Web Services* – As a search engine serves requests to look for interesting compositions of Web services, a service mining engine can tap into information gathered there to formulate some initial mining contexts and spawn off relevant mining threads. Alternatively, it can start such mining activities with little or no input from a human in the loop. In this case, the weight of the evaluation of interestingness and usefulness will be shifted further downstream. We discussed the subjective nature of both of these two measures earlier in the book. If a great deal of mining results are captured and stored, then the subjective evaluation of these two measures can be delayed until an end user is later presented with a relevant set of results by a mining service on the Web. It is conceivable that different individuals will in some cases be gravitated towards some common results due to their universal appeals and hone in different results in others due to their more subjective appeals. A service-based mining approach will need to address the provisioning of mining context and the balancing of the conflicting objectives to identify universally appealing results early on and to store as much mining results as possible for delayed subjective evaluation.

*Elaborate Service Modeling Methodologies* – In this research, we focused on using three types of service/operation recognition mechanisms to model interactions among biological entities. These include indirect recognition, promotion and inhibition. In reality, interactions among biological entities can be very diverse and more complicated. For example, BioPAX Level 1 [18] defines the following interactions:

- *Conversion* – An interaction in which one or more entities is physically transformed into one or more other entities. *Biochemical reaction* is a conversion interaction in which one or more entities (substrates) undergo covalent changes to become one or more other entities (products). *Complex assembly* is a conversion interaction in which a set of physical entities, at least one being a macromolecule (protein or RNA), aggregate via non-covalent interactions.
- *Control* – An interaction in which one entity regulates, modifies, or otherwise influences another. *Catalysis* is a control interaction in which a physical entity (a catalyst) increases the rate of a conversion interaction by lowering its activation energy. *Modulation* is a control interaction in which a physical entity modulates a catalysis interaction.

To accurately capture the details of biological interactions, each of the above needs to be sufficiently represented using the Web service modeling paradigm. Obviously, domain-specific research is needed to categorize the behavioral varieties of each application domains. As new modeling mechanisms arise, service mining strategies outlined in this research will then need to be adapted accordingly.

# References


1. Apache axis2/java - next generation web services. http://ws.apache.org/axis2/.
2. Bps: Biochemical pathway simulator. http://www.brc.dcs.gla.ac.uk/projects/bps/.
3. Copasi. http://mendes.vbi.vt.edu/tiki-index.php?page=COPASI.
4. The graphml file format. http://graphml.graphdrawing.org/.
5. Java architecture for xml binding, binding compiler (xjc). http://java.sun.com/webservices/docs/1.6/jaxb/xjc.html.
6. Jetty. http://www.mortbay.org/.
7. Nf-kappab pathway. http://www.cellsignal.com/reference/pathway/NF_kappaB.jsp.
8. Resource description framework (rdf). http://www.w3.org/RDF/.
9. Uniprotkb/swiss-prot. http://www.ebi.ac.uk/swissprot/.
10. The web service modeling language wsml. http://www.wsmo.org/wsml/wsml-syntax.
11. Web service modeling ontology. http://www.wsmo.org/.
12. The web service modeling toolkit wsmt. http://sourceforge.net/projects/wsmt.
13. Web services description language (wsdl) 1.1. http://www.w3.org/TR/wsdl.
14. Web services execution environment. http://sourceforge.net/projects/wsmx.
15. Web Services Flow Language (WSFL). Technical report, IBM. http://xml.coverpages.org/wsfl.html.
16. Web services semantics - wsdl-s. http://www.w3.org/Submission/WSDL-S/.
17. yed - java graph editor. http://www.yworks.com/en/products_yed_about.htm.
18. Biopax - biological pathways exchange language, level 1, version 1.2 documentation, biopax recommendation. Technical report, BioPAX Workgroup, November 2004. http://www.biopax.org.
19. Owl-s: Semantic markup for web services. November 2004. http://www.w3.org/Submission/2004/SUBM-OWL-S-20041122/.
20. Web services architecture - w3c working group note. Technical report, February 2004. http://www.w3.org/TR/2004/NOTE-ws-arch-20040211/.
21. ActiveBPEL. Activebpel engine. http://www.activebpel.org/.
22. J. Augen. The evolving role of information technology in the drug discovery process. *Drug Discovery Today*, 7:315–323, 2002.
23. Sunny Y. Auyang. From experience to design - the science behind aspirin. http://www.creatingtechnology.org/biomed/aspirin.htm.
24. Thomas Baeck, D. Fogel, and Z. Michalewicz (editors). *Handbook of Evolutionary Computation*. IOP Publishing, UK, 1997.
25. Philip Ball. *Designing the Molecular World - Chemistry at the Frontier*. Princeton University Press, Princeton, New Jersey, 1994.
26. BEA, IBM, and Microsoft. Business process execution language for web services (bpel4ws). http://xml.coverpages.org/bpel4ws.html.

27. Alex Berson, Kurt Thearling, and Stephen J. Smith. *Building Data Mining Applications for CRM*. McGraw-Hill, December 1999.

28. Upinder S. Bhalla and Ravi Iyengar. Emergent properties of networks of biological signaling pathways. *Science*, 283:381 – 387, 1999.

29. Stephan Borzsonyi, Donald Kossmann, and Konrad Stocker. The skyline operator. In *Proceedings of the 17th International Conference on Data Engineering*, pages 421 – 430, Washington, DC, 2001.

30. Roger Brent and Jehoshua Bruck. Can computers help to explain biology? *Nature*, 440(23):416 – 417, March 2006.

31. Luca Cardelli. Abstract machines of systems biology. 3737:145–168, 2005.

32. Jacques Cohen. Bioinformatics: An introduction for computer scientists. *ACM Computing Surveys*, 36(2):122 – 158, 2004.

33. C. Coves, D. Crestani, and F. Prunet. Design and analysis of workflow processes with petri nets. *IEEE International Conference on Systems, Man, and Cybernetics*, 1:101–106, 1998.

34. Charles Darwin. *The Origin of Species by Means of Natural Selection*. Adamant Media Corporation, 2005. Original 1859.

35. Leandro Nunes de Castro and Jonathan Timmis. *Artificial Immune Systems: A New Computational Intelligence Approach*. Springer, 2001.

36. Hidde de Jong and Michel Page. Qualitative simulation of large and complex genetic regulatory systems. In *Proceedings of the 14th European Conference on Artificial Intelligence, Berlin, Germany, August 20-25*, pages 141–145, 2000.

37. W. Derks, J. Dehnert, P. Grefen, and W. Jonker. Customized atomicity specification for transactional workflows. *The Proceedings of the Third International Symposium on Cooperative Database Systems for Advanced Applications, CODAS2001*, pages 140–147, 2001.

38. M. Dorigo. *Optimization, Learning and Natural Algorithms*. PhD thesis, Politecnico di Milano, 1992.

39. Sorin Drăghici and R. Brian Potter. Predicting hiv drug resistance with neural networks. *Bioinformatics*, 19(1):98 – 107, 2003.

40. S. Dustdar, T. Hoffmann, and W. van der Aalst. Mining of ad-hoc business processes with teamLog, 2005. Data and Knowledge Engineering, 2005.

41. Andries Engelbrecht. *Fundamentals of Computational Swarm Intelligence*. Wiley, 2006.

42. Francisco Curbera et al. Business process execution language for web services, version 1.0. http://www.ibm.com/developerworks/webservices/library/ws-bpel/, July 2002.

43. J. Farmer, N. Packard, and A. Perelson. The immune system, adaptation, and machine learning. *Physica D*, 22:187–204, 1986.

44. Adam Farquhar, Richard Fikes, and James Rice. The ontolingua server: a tool for collaborative ontology construction. *International Journal of Human-Computer Studies*, 46:707–727, 1997.

45. Craig Freudenrich. How pain works. http://health.howstuffworks.com/pain.htm.

46. K. Fukushima. A self-organizing multilayered neural network. In *Biological Cybernetics*, volume 20, pages 121–136, 1975.

47. W.E. Grosso, H. Eriksson, R.W. Fergerson, J.H Gennari, S.W. Tu, and M.A. Musen. Knowledge modeling at the millennium (the design and evolution of protege-2000i). *Twelfth Banff Workshop on Knowledge Acquisition, Modeling, and Management*, 1999.

48. Rachid Hamadi and Boualem Benatallah. A petri net-based model for web service composition. In *CRPITS'17: Proceedings of the Fourteenth Australasian database conference on Database technologies 2003*, pages 191–200, Darlinghurst, Australia, Australia, 2003. Australian Computer Society, Inc.

49. Jiawei Han and Micheline Kamber. *Data Mining: Concepts and Techniques*. Morgan Kaufmann Publishers, 2000.

50. Jim Hendler. A brief history of web ontology work. http://www.w3.org/2002/Talks/www2002-ont-jh/slide7-0.html.

51. Javier Herrero, Alfonso Valencia, and Joaquin Dopazo. A hierarchical unsupervised growing neural network for clustering gen expression patterns. *Bioinformatics*, 17(2):126 – 136, 2001.

52. Lucas Hoffman. How aspirin works. http://health.howstuffworks.com/aspirin1.htm.

53. Horst Ibelgaufts. Cope - cytokines online pathfinder encyclopaedia. http://www. copewith-cytokines.de/.

54. IBM. Websphere software. http://www-306.ibm.com/software/websphere/.

55. Trey Ideker, Timothy Galitski, and Leroy Hood. A new approach to decoding life: Systems biology. *Annual Review of Genomics and Human Genetics*, 2:343–371, 2001.

56. John Hagel III and John Seely Brown. Your next it strategy. *Harvard Business Review*, pages 105 – 113, October 2001.

57. Stat Soft Inc. Data mining techniques. http://www.uprforum.com/index.html, 2003.

58. S. Istrail, S. Ben-Tabou De-Leon, and E. Davidson. The regulatory ggenome and the computer. *Developmental Biology*, 310:187–195, 2007.

59. M. Kanehisa, S. Goto, M. Hattori, K. F. Aoki-Kinoshita, M. Itoh, S. Kawashima, T. Katayama, M. arki, and M. Hirakawa. From genomics to chemical genomics: new developments in kegg. *Mucleic Acids Research*, 34:354 – 357, 2006.

60. Lila Kari and Grzegorz Rozenberg. The many facets of natural computing. *Communications of the ACM*, 51(10):72–83, 2008.

61. Peter D. Karp, Suzanne Paley, and Pedro Romero. The pathway tools software. *Bioinformatics*, 18:S1 – S8, 2002.

62. T. Kohonen. *Self-Organizing Maps*, volume 30. Springer, 1995.

63. Kanehisa Laboratories. Kegg: Kyoto encyclopedia of genes and genomes. http://www. genome.jp/kegg/.

64. Misia Landau. Inflammatory villain turns do-gooder. http://focus.hms.harvard.edu/ 2001/Aug10_2001/immunology.html.

65. Frank Leymann. Web Services Flow Language (WSFL 1.0). Technical report, IBM, May 2001.

66. Sea Ling and Heinz Schmidt. Time petri nets for workflow modelling and analysis. *IEEE International Conference on Systems, Man, and Cybernetics*, 4:3039–3044, 2000.

67. Seth Lloyd. *Programming the Universe: A Quantum Computer Scientist Takes on the Cosmos*. Knopf, 2006.

68. W. McCulloch and W. Pitts. A logical calculus of the ideas immanent in nervous activity. *Bulletin of Mathematical Biophysics*, 5:115–133, 1943.

69. Daniel M. McDonald, Hsinchun Chen, Hua Su, and Byron B. Marshall. Extractin gene pathway relations using a hybrid grammar: the arizona relation parser. *Bioinformatics*, 20(182004):3370 – 3378, July 2004.

70. Brahim Medjahed. *Semantic Web Enabled Composition of Web Services*. PhD thesis, Department of Computer Science and Applications, Virginia Polytechnic Institute and State University, Falls Church, VA, 2004.

71. Brahim Medjahed, Athman Bouguettaya, and Ahmed K. Elmagarmid. Composing web services on the semantic web. *The VLDB Journal*, September 2003.

72. L. G. Meredith and Steve Bjorg. Contracts and types. *Commun. ACM*, 46(10):41–47, 2003.

73. Drug Design Methodologies. Theory of drug design. http://www.newdrugdesign.com/ Rachel_Theory_01.html.

74. Microsoft. Microsoft biztalk server. http://www.microsoft.com/biztalk.

75. R. Milner. The polyadic $\pi$-calculus: A tutorial. *Logic and Algebra of Specification*, pages 203 – 246, 1993.

76. NCBI. Genbank. http://www.ncbi.nlm.nih.gov/Genbank/.

77. See-Kiong Ng and Marie Wong. Toward routine automatic pathway discovery from on-line scientific text abstracts. *Genome Informatics*, 10:104 – 112, 1999.

78. OASIS. Business process modeling language. http://xml.coverpages.org/bpml.html.

79. OASIS. Business transaction protocol. http://xml.coverpages.org/ni2004-12-24-a.html, 2002.

80. Adaptive Computing Laboratories of Iowa. Electric load forecacsting methods. http://www.aclillc.com/.

81. OpenStorm. Openstorm service orchestrator. http://www.openstorm.com/.

82. Oracle. Oracle bpel process manager. http://www.oracle.com/technology/products/ias/bpel/index.html.

83. Linus Pauling. *General Chemistry*. Dover Publications, Inc., Mineola, New York, 1988.

84. C. Reynolds. Flocks, herds, and schools: A distributed behavioral model. *Computer Graphics*, 21(4):25–34, 1987.

85. F. Rosenblatt. *Principles of Neurodynamics*. Spartan, 1962.

86. D. Rumelhart, G. Hinton, and R. Williams. Learning internal representations by error propagation. In *Parallel Distributed Processing, Explorations in the Microstructure of Cognitron*, volume 1, Cambridge, MA, 1986. MIT press.

87. Carlos Santos, Daniela Eggle, and David J. States. Wnt pathway curation using automated natural language processing: combining statistical methods with partial and full parse for knowledge extraction. *Bioinformatics*, 21(82005):1653 – 1658, November 2005.

88. Munindar P. Singh and Michael N. Huhns. *Service-Oriented Computing: Semantics, Processes, Agents*. John Wiley & Sons, 2005.

89. Nicholas Snowden. Nucleic acids: Structures and functions. http://www.nicksnowden.net/Module_1_pages/nucleic_acids%20and%20protein_synthesis.htm.

90. Biplav Srivastava and Jana Koehler. Web service composition - current solutions and open problems. In *ICAPS 2003 Workshop on Planning for Web Services*, pages 28–35, Trento, Italy, June 2003.

91. Gary Stix. Molecular treasure hunt. *Scientific American*, pages 88 – 91, May 2005.

92. Satish Thatte. XLANG - Web Services For Business Process Design. Technical report, Microsoft, 2001. http://www.gotdotnet.com/team/xml_wsspecs/xlang-c/default.htm.

93. Masaru Tomita, Kenta Hashimoto, Koichi Takahashi, Thomas Simon Shimizu, Yuri Matsuzaki, Fumihiko Miyoshi, K. Saito, S. Tanida, Katsuyuki Yugi, J. C. Venter, and C. A. Hutchison III. E-cell: Software environment for whole-cell simulation. *Bioinformatics*, 15(1):72 – 84, 1999.

94. UCLA. Database of interacting proteins. http://dip.doe-mbi.ucla.edu/.

95. UN/CEFACT and OASIS. Electronic business xml initiative (ebxml). http://xml.coverpages.org/ebXML.html.

96. W.M.P van der Aalst. Generic workflow models: how to handle dynamic change and capture management information? *Proceedings of 1999 IFCIS International Conference on Cooperative Information Systems*, pages 115–126, 1999.

97. W3C. Web service choreography interface (wsci) 1.0. http://www.w3.org/TR/wsci/.

98. W3C. Web services architecture. http://www.w3.org/TR/2002/WD-ws-arch-20021114/.

99. W3C. http://www.daml.org/services/owl-s/. 2004.

100. Rick Whiting. Tower of power. *Information Week*, February 2002.

101. Tim Berners-Lee with Mark Fischetti. *Weaving the Web - The Original Design and Ultimate Destiny of the World wide Web by Its Inventor*. HarperSanFrancisco, 1999.

102. Ian H. Witten and Eibe Frank. *Data Mining*. Morgan Kaufmann Publishers, San Francisco, California, 2000.

103. Jun Xu and Arnold Hagler. Chemoinformatics and drug discovery. *Molecules*, 7:566–600, 2002.

104. Daming Yao, Jingbo Wang, Yanmei Lu, Nathan Noble, Huandong Sun, Xiaoyan Zhu, Nan Lin, Donald G. Payan, Ming Li, and Kunbin Qu. Pathwayfinder: Paving the way toward automatic pathway extraction. In *Proceedings of the Second Conference on Asia-Pacific Bioinformatics*, volume 29, pages 52 – 62, Dunedin, New Zealand, 2004. Australian Computer Society, Inc.

105. Min-Jean Yin, Yumi Yamamto, and Richard B. Gaynor. The anti-inflammatory agents aspirin and salicylate inhibit the activity of I$\kappa$B kinase-$\beta$. *Nature*, 369:77 – 80, November 1998.

106. Qi Yu. *A Foundational Framework for Service Query Optimization*. PhD thesis, Department of Computer Science and Applications, Virginia Polytechnic Institute and State University, Blacksburg, VA, 2008.

107. Liangzhao Zeng, B. Benatallah, A. H. H. Ngu, M. Dumas, J. Kalagnanam, and H. Chang. Qos-aware middleware for web services composition. *IEEE Transactions on Software Engineering*, 30(5):311–327, 2004.

108. George Zheng and Athman Bouguettaya. Mining web services for pathway discovery. In *2007 VLDB Workshop on Data Mining in Bioinformatics*, Vienna, Austria, September 2007.

109. George Zheng and Athman Bouguettaya. A web service mining framework. In *Proceedings of IEEE International Conference on Web Services (ICWS)*, pages 1096–1103, Salt Lake City, Utah, USA, July 2007. IEEE Computer Society.

110. George Zheng and Athman Bouguettaya. Discovering pathways of service oriented biological processes. In James Bailey, David Maier, Klaus-Dieter Schewe, Bernhard Thalheim, and Xiaoyang Sean Wang, editors, *Proceedings of the 9th International Conference on Web Information Systems Engineering (WISE)*, volume 5175 of *Lecture Notes in Computer Science*, pages 189–205, Auckland, New Zealand, September 2008. Springer.

111. George Zheng and Athman Bouguettaya. Service-based analysis of biological pathways. *BMC Bioinformatics, Supplement on Semantic Web Applications and Tools for LIfe Sciences*, 10(10), October 2009.

112. George Zheng and Athman Bouguettaya. Service mining on the web. *IEEE Transactions on Services Computing*, 2(1):65–78, 2009.

# Glossary

Use the template *glossary.tex* together with the Springer document class SVMono (monograph-type books) or SVMult (edited books) to style your glossary in the Springer layout.

**AI**  Artificial Intelligence

**API**  Application Programming Interface

**ATP**  Adenosine Triphosphate, a high energy phosphate compound

**BPEL4WS**  Business Process Execution Language for Web Services

**BPML**  Business Process Modeling Language

**BTP**  Business Transaction Protocol

**COPE**  Cytokines Online Pathfinder Encyclopedia

**CORBA**  Common Object Request Broker Architecture

**COX**  Cyclooxygenase

**COX1**  Cyclooxygenase-1 or Prostaglandin-endoperoxide synthase 1

**COX2**  Cyclooxygenase-2 or Cytochrome c oxidase subunit II

**CWS**  Conference Web Service

**DAML+OIL**  DARPA Agent Markup Language Ontology Inference Layer

**DAML-S**  DARPA Agent Markup Language for Services

**DARPA**  Defense Advanced Research Projects Agency

**DIP**  Database of Interacting Proteins

**DNA**  Deoxyribonucleic Acid

**ebXML**  Electronic Business XML

**EDI**  Electronic Data Interchange

**EJB**  Enterprise Java Beans

**FD**  Functional Description

**GraphML**  Graph Markup Language

**GRL**  Group of Relevant Leads

**I$\kappa$B**  Inhibitory kappa B

**KEGG**  Kyoto Encyclopedia of Genes and Genomes

**LTB4**  Leukotriene B4

**LXA4**  Lipoxin A4

**mRNA**  Messenger RNA

**NF-$\kappa$B/Rel**  Nuclear Factor kappa B

**NFP**  Non Functional Property

**NLP**  Natural Language Processing

**OASIS**  Organization for the Advancement of Structured Information Standards

**OWL-S**  Web Ontology Language for Services

**PG**  Prostaglandin

**PGE2**  Prostaglandin E2

**PGG2**  Prostaglandin G2

**PGI2**  Prostaglandin I2

**PLA2**  Phospholipase A2

**QoWS**  Quality of Web Service

**RDF**  Resource Description Framework

**RNA**  Ribonucleic Acid

**SOAP**  Simple Object Access Protocol

**SQL**  Structured Query Language

**Swiss-Prot**  A protein knowledgebase maintained by Swiss Institute for Bioinformatics

**tRNA**  Transfer RNA

**UDDI**  Universal Description, Discovery, and Integration

**UML**  Unified Modeling Language

**W3C**  World Wide Web Consortium

**WSCI**  Web Service Choreography Interface

**WSD**  World Service Description

**WSDL**  Web Service Description Language

**WSDL-S**  WSDL with Semantics

**WSFL**  Web Services Flow Language

**WSML**  Web Service Modeling Language

**WSMO**  Web Service Modeling Ontology

**WSMT**  Web Service Modeling Toolkit

**WSMX**  Web Service Modeling eXecution environment

**XLANG**  An XML-based extension of WSDL

**XML**  Extensible Markup Language

# Index